Simplified Quantum Computing with Applications

IOP Series in Coherent Sources, Quantum Fundamentals, and Applications

About the Editor

F J Duarte is a laser physicist based in Western New York, USA. His career has expanded three continents while contributing in the academic, industrial and defense sectors. Duarte is editor/author of 15 laser optics books and sole author of three books: *Tunable Laser Optics, Quantum Optics for Engineers*, and *Fundamentals of Quantum Entanglement*. Duarte has made original contributions in the fields of coherent imaging, directed energy, high-power tunable lasers, laser metrology, liquid and solid-state organic gain media, narrow-linewidth tunable laser oscillators, organic semiconductor coherent emission, *N*-slit quantum interferometry, polarization rotation, quantum entanglement, and space-to-space secure interferometric communications. He is also the author of the generalized multiple-prism grating dispersion theory and pioneered the use of Dirac's quantum notation in *N*-slit interferometry and classical optics. His contributions have found applications in numerous fields, including astronomical instrumentation, dispersive optics, femtosecond laser microscopy, geodesics, gravitational lensing, heat transfer, laser isotope separation, laser medicine, laser pulse compression, laser spectroscopy, mathematical transforms, nonlinear optics, polarization optics, and tunable diode laser design. Duarte was elected Fellow of the Australian Institute of Physics in 1987 and Fellow of the Optical Society of America in 1993. He has received various recognitions, including the *Paul F Foreman Engineering Excellence Award* and the *David Richardson Medal* from the Optical Society.

Coherent Sources, Quantum Fundamentals, and Applications

Since its discovery, the laser, has found innumerable applications from astronomy to… zoology. Subsequently, we have also become familiar with additional sources of coherent radiation such as the free electron laser, optical parametric oscillators, and coherent interferometric emitters. The aim of this book Series in Coherent Sources, Quantum Fundamentals, and Applications is to explore and explain the physics and technology of widely applied sources of coherent radiation and to match them with utilitarian and cutting-edge scientific applications. Coherent sources of interest are those that offer advantages in particular emission characteristics areas such as broad tunability, high spectral coherence, high energy, or high power. An additional area of inclusion are those coherent sources capable of high performance in the miniaturized realm. Understanding of quantum fundamentals can lead to new and better coherent sources and unimagined scientific and technological applications. Application areas of interest include the industrial, commercial, and medical sectors. Also, particular attention is given to scientific applications with a bright future such as coherent spectroscopy, astronomy, biophotonics, space communications, space interferometry, quantum entanglement, and quantum interference.

Publishing benefits
Authors are encouraged to take advantage of the features made possible by electronic publication to enhance the reader experience through the use of colour, animation and video, and incorporating supplementary files in their work.

Do you have an idea of a book that you'd like to explore?
For further information and details of submitting book proposals, see iopscience.org/books or contact Ashley Gasque at ashley.gasque@iop.org.

A full list of titles published in this series can be found here: https://iopscience.iop.org/bookListInfo/series-in-coherent-sources-and-applications.

Simplified Quantum Computing with Applications

Koji Nagata
Department of Physics, Korea Advanced Institute of Science and Technology, Daejeon, Republic of Korea. E-mail: ko_mi_na@yahoo.co.jp

Do Ngoc Diep
TIMAS, Thang Long University, Nghiem Xuan Yem Road, Hoang Mai District, Hanoi, Vietnam
and
Institute of Mathematics, VAST, 18 Hoang Quoc Viet Road, Cau Giay District, Hanoi, Vietnam. E-mail: dndiep@math.ac.vn

Ahmed Farouk
Department of Computer Science, Faculty of Computers and Artificial Intelligence, South Valley University, Hurghada, Egypt. E-mail: ahmed.farouk@sci.svu.edu.eg

Tadao Nakamura
Department of Information and Computer Science, Keio University, 3-14-1 Hiyoshi, Kohoku-ku, Yokohama 223-8522, Japan. E-mail: nakamura@pipelining.jp

IOP Publishing, Bristol, UK

ISBN 978-0-7503-4700-6 (ebook)
ISBN 978-0-7503-4698-6 (print)
ISBN 978-0-7503-4701-3 (myPrint)
ISBN 978-0-7503-4699-3 (mobi)

DOI 10.1088/978-0-7503-4700-6

Version: 20220701

IOP ebooks

British Library Cataloguing-in-Publication Data: A catalogue record for this book is available from the British Library.

Published by IOP Publishing, wholly owned by The Institute of Physics, London

IOP Publishing, Temple Circus, Temple Way, Bristol, BS1 6HG, UK

US Office: IOP Publishing, Inc., 190 North Independence Mall West, Suite 601, Philadelphia, PA 19106, USA

Contents

Preface

Until recently quantum computing was not always confined to being very true itself in terms of the technology of quantum applied computers as quantum-gated computers. Of course, the other technological quantum-oriented annealing style computers have been productive and businesslike machines. However, very true quantum computers are not such annealers but quantum-gated computers in terms of the notion of computing.

This book focuses on the fundamental concepts of quantum-gated computing from the beginning: the Deutsch function evaluation algorithm, the Deutsch–Jozsa algorithm, the Grover's search algorithm, the Bernstein–Vazirani algorithm for identifying linear functions, the Simon period finding algorithm, the Simon hidden subgroup algorithm and the Shor prime factoring algorithm. With computation in mind, this book takes a close look at the extraordinary connection between the Boolean algebra and quantum-gated computing as well as the new applications in a simplified manner. Our sincere objective is to build up quantum-gated computers operated by existing software/programs for electronic computers. Therefore, before doing this we study all the fundamental concepts on quantum-gated computers.

Acknowledgements

We would like to thank all our colleagues, especially Soliman Abdalla, Jaewook Ahn, Josep Batle, Mark Behzad Doost, Han Geurdes, Preston Guynn, Shahrokh Heidari, Wenliang Jin, Hamed Daei Kasmaei, Janusz Milek, Mosayeb Naseri, Santanu Kumar Patro, Germano Resconi, and Renata Wong for their valuable support at our original paper publication level. On the other hand, we cannot forget all the efforts of those who helped us in the preparation for publishing this book. In particular, Senior Commissioning Editor Ashley Gasque, Ebooks Editorial Assistants Claire Roy and Robert Trevelyan, and Accounts Payable Assistant Sam Needs deserve our sincere thanks for their complicated paperwork toward the great hard work in publishing the book.

Author biographies

Koji Nagata

Koji Nagata received the BS and MS degrees from Kyoto University and Tohoku University in 1996 and 2000, respectively. He graduated with a PhD from The Graduate University of Advanced Sciences' PhD (School of Advanced Sciences) course in March 2003. His academic work has recently been broadly based on aspects of quantum information theory. His professional experiences are as follows: Research fellow at National Institute of Information and Communications Technology (NICT), Guest research fellow at Institute of Theoretical Physics and Astrophysics, University of Gdansk, Research Professor (and Research fellow) at Department of Physics, Korea Advanced Institute of Science and Technology (KAIST), Research fellow at Obihiro University of Agriculture and Veterinary Medicine, and further Research fellow at Future University Hakodate.

Do Ngoc Diep

Do Ngoc Diep, graduated from Lomonosov Moscow State University in 1975 and there received a PhD degree in 1977. In 1978 he started work at Institute of Mathematics, Vietnam Academy of Science and Technology. In 1995 he received the DSc degree (Habilitation). In 1996 he became a full professor. He was a senior researcher, Head of the Department of Geometry and Topology, and Department of Mathematical Analysis of the Institute of Mathematics, VAST. In 2017 he moved to Thang Long Institute of Mathematics and Applied Sciences, Thang Long University.

His main fields of research are: methods of noncommutative geometry for group C*-algebras, KK-functor invariants of group C*-algebras and quantum groups, multidimensional quantization and quantum computing, quantum neural networks.

He has published more than 80 research papers.

Published monograph and books: 1. Do Ngoc Diep 1999 *Noncommutative Geometry Methods for Group C*-algebras*, X (Research Notes in Mathematics vol. 416) (Boca Raton, FL, New York London etc: Chapman and Hall); 2. Do Ngoc Diep and Nong Quoc Chinh 2010 *Lectures on Differential Geometry, Graduate course* (Hanoi: VNU publishing); 3. Book chapter 5 'Quantum Finance' in the book 2022 *Quantum Computing environment* (Berlin: Springer Nature).

Ahmed Farouk

Ahmed Farouk is currently assistant professor at South Valley University. Before that he was a Postdoctoral Research Fellow at Wilfrid Laurier University and Ryerson University, Canada. He received his MSc and PhD degrees from Mansoura University, Egypt. He is one of the top 20 technical co-founders of the Quantum Machine Learning Program by Creative Destruction Lab at the University of Toronto. Furthermore, he is selected as Top 25 of InnovateTO150 Canada to showcase the best of Toronto's next generation of change-makers, innovators, and entrepreneurs. He is exceptionally well known for his seminal contributions to theories of quantum information, machine learning, and cryptography. He has published over 80 papers in reputed and high impact journals like *IEEE Internet of Things, IEEE Transactions on Intelligent Transportation Systems, Nature Scientific Reports and Physical Review* A. The exceptional quality of his research is recognized nationally and internationally. He was selected by the scientific review panel of the Council for the Lindau Nobel Laureate Meetings to participate in the 70th Lindau Nobel Laureate Meeting. His volunteering work is apparent since he was appointed as chair of the IEEE computer chapter for the Waterloo-Kitchener area and editorial board for many reputed journals like *Nature Scientific Reports, IET Quantum Communication*, and *IEEE Access*. Also, he was selected for IEEE and IET Young Professional Ambassador and as a moderator for the new IEEE TechRxiv. Recently, he was appointed as an associate editor for the IEEE Canadian Review (ICR), deputy editor for *Elsevier Neuroscience Informatics* journal and academic editor for Hindawi *Mobile Information Systems Journal*.

Tadao Nakamura

Tadao Nakamura is currently a Professor Emeritus of Tohoku University, and also a Professor (as a visiting status) of Keio University. In 1994 he was given the status of a Full Professor (as a visiting status) in the Electrical Engineering Department at Stanford University. And even today he still stays at Stanford at any time. In 2007 he was also induced as a Professorial Fellow at Imperial College London. His research interests are toward electronic and quantum-gated computer systems. In 2004 he received the IEEE Computer Society's Taylor L. Booth Award. He has been Steering Committee Chair after the Organizing and Advisory Committee Chairs of COOL Chips conference series fully sponsored by the IEEE Computer Society. He is a Life Fellow of the IEEE.

Abstract

The objectives of this book are not only for beginners to learn quantum-gated computers but also the experts to develop them much more toward the very true quantum-gated ones. We present a secure quantum key distribution based on a special Deutsch–Jozsa algorithm using Greenberger–Horne–Zeilinger states, discuss the application of the Bernstein–Vazirani algorithm to quantum communication including error correction and also study the Bernstein–Vazirani algorithm in a noisy environment. We propose methods for evaluating many functions and for computing a multiplication operation, both of which use the generalized Bernstein–Vazirani algorithm together with many parallel quantum systems. We study quantum communication based on Simon's algorithm and generalize Deutsch's algorithm for determining all the mappings of a function. Proposing a parallel computation for all of the combinations of values in variables of a logical function, we demonstrate some parallel evaluations of two two-variable functions which are elements of a Boolean algebra composed of the four-atom set. Using quantum cryptography based on an algorithm for determining a function using qudit systems, we propose continuous-variable quantum computing and its applications to cryptography. We present various new forms of the Bernstein–Vazirani algorithm beyond qubit systems. We search an algorithm for rapidly determining a homogeneous linear function, construct quantum algorithms for finding the roots of a polynomial function and suggest a quantum algorithm for rapidly plotting a function. Considering an algorithm that determines whether or not a given function is even, we propose a necessary and sufficient condition for quantum computing. Finally, studying our problem toward practical quantum-gated computers, we propose the concrete quantum circuits for addition of any two numbers and we study its computational complexity. Additionally, we study the measurement theory based on the truth values.

IOP Publishing

Simplified Quantum Computing with Applications

Koji Nagata, Do Ngoc Diep, Ahmed Farouk and Tadao Nakamura

Chapter 1

Introduction

This chapter provides a general introduction and historical background of this work. Some references concerning the work are given. We provide the contents of the work.

1.1 Introduction

Quantum mechanics is successful in explaining and predicting many physical phenomena [1–6], and there is therefore good reason to believe in its correctness. Among the interesting applications of quantum principles is their application to information theory [6], which leads to the concept of quantum computing.

Quantum computing is explained as follows [7]: the design and theory of computer systems that depend on quantum effects for their operation. On one level, this can be the use of small components, at the atomic or molecular level, to store or process information. An example would be a storage system that used two different spin states of atoms to store bits of information, or a logic gate that depends on the movement or spin of a single electron. Systems of this type are studied in nanocomputing. At a more fundamental level, the term 'quantum computing' implies the use of quantum effects that have no classical analogue to process information. In a 'classical' computer information is held in bits, which can have two alternative values (0 and 1). In a quantum computer the 0 and 1 values are held simultaneously in a superposition state. This unit of information is called a quantum bit (or qubit). Much more information can be held in this way and, in principle, it is possible to do parallel processing of the information. Quantum computers would be much faster than conventional machines and capable of performing calculations that could not realistically be done otherwise. Ion traps, cavity QED, and spin measurements have been used in research in this area.

Articles of research for constructing theoretical quantum algorithms [7] may be mentioned as follows. In 1985, the Deutsch algorithm was introduced and constructed for the function property problem [8–10]. In 1993, the Bernstein–Vazirani algorithm was proposed for identifying linear functions [11, 12]. More recently, a generalized

doi:10.1088/978-0-7503-4700-6ch1

Bernstein–Vazirani algorithm has been investigated [13]. In 1994, Simon's algorithm [14] and Shor's algorithm [15] were discussed for period finding of periodic functions. In 1996, Grover [16] provided an algorithm for unordered object finding and the motivation for exploring the computational possibilities offered by quantum mechanics.

The implementation of a quantum algorithm to solve Deutsch's problem on a nuclear magnetic resonance quantum computer was first reported by Jones and Mosca [17]. An implementation of the Deutsch–Jozsa algorithm on an ion-trap quantum computer was also reported [18]. There were several attempts to use single-photon two-qubit states for quantum computing. Oliveira *et al* [19] implemented Deutsch's algorithm with polarization and transverse spatial modes of the electromagnetic field as qubits. In addition, single-photon Bell states were prepared and measured [20]. A decoherence-free implementation of Deutsch's algorithm was achieved using such single-photon states and two logical qubits [21]. A one-way-based experimental implementation of Deutsch's algorithm was reported [22].

The implementation of a quantum algorithm to solve the Bernstein–Vazirani parity problem without entanglement in an ensemble quantum computer is another important development [23]. Fiber-optics implementations of the Deutsch–Jozsa and Bernstein–Vazirani quantum algorithms with three qubits were proposed [24]. Efficient optical implementation of the Bernstein–Vazirani algorithm was reported [25].

Other developments in recent years include the following. In real experiments, it is impossible to avoid the effects of environmental noise, and the question of whether or not quantum learning is robust against noise is a subject of intensive study [26]. A quantum algorithm for approximating the influence of Boolean functions and its applications were studied by Li and Yang [27]. Quantum computation with coherent spin states and the close Hadamard problem were reported by Adcock *et al* [28]. A transport implementation of the Bernstein–Vazirani algorithm with ion qubits was studied by Fallek *et al* [29]. In 2016, Krishna *et al* introduced a generalization of the Bernstein–Vazirani algorithm to qudit systems [30]. A simple algorithm for complete factorization of an N-partite pure quantum state was discussed [31]. Fujikawa *et al* discussed a classical limit of Grover's algorithm induced by dephasing: coherence versus entanglement [32]. Quantum dialogue protocol based on Grover's search algorithms was presented [33].

Quantum Gauss–Jordan elimination and simulation of accounting principles on quantum computers were discussed by Diep *et al* [34]. A dynamical analysis of Grover's search algorithm in arbitrarily high-dimensional search spaces was given by Jin [35]. A method of computing many functions simultaneously by using multiple parallel quantum systems was reported by Nagata *et al* [36]. Scalable networks-on-chip with elastic links demarcated by decentralized routers was studied by Yasudo *et al* [37].

The double-slit experiment [38, 39] is an illustration of wave–particle duality—a concept that many find hard to understand. Let us consider the double-slit experiment from the point of view of energy. In this experiment, a source emits a beam of energy in the form of photons, which are particles. The beam travels through a barrier in which there are two slits. If one places a detector screen on the other side of the barrier, the pattern of detected energy exhibits interference fringes, which are characteristic of waves, although the detector actually responds to the energy of particles. Thus, the system exhibits behavior that although it can be interpreted both in terms of waves (interference patterns) and in terms of particles (dots on the screen), is basically related to energy. In this energy interpretation, a quantum state is a distribution of energy. Thus, quantum measurement theory involves investigating distributions of energy. This approach allows us to create genuine quantum algorithms.

An algorithm for fast determining a homogeneous linear function is proposed [40]. A method of calculating a multiplication by using the generalized Bernstein–Vazirani algorithm is studied [41]. A new mathematical structure for quantum algorithms in the case of a special function is reported [42]. Standard quantum algorithms and the property of a certain function are reported [43]. Efficient quantum algorithm for the parity problem of a certain function is given [44]. The necessary and sufficient condition for quantum computing is proposed [45].

The argumentations for finding this evenness: $f(x) = f(-x)$ of the function $f(x)$ is assumed to a practical use in combination together with quantum error correction algorithms for qubits/qudits (or even topological anyons in topological quantum error correction), see [6] and [46]. After an error correction the problem is protected from any kinds of bit flip or spin flip actions and the argumentations work well.

Efficient quantum algorithms of finding the roots of a polynomial function are discussed [47, 48]. Designing a quantum circuit by one-step method and similarity with a neural network are discussed [49]. A quantum algorithm for determining a complex number string is reported [50]. Generalization of the Bernstein–Vazirani algorithm beyond qubit systems is reported [51]. Various new forms of the Bernstein–Vazirani algorithm beyond qubit systems are studied [52]. A parallel computation for all of the combinations of values in variables of a logical function is proposed [53]. We discuss the relation between set theory [54] based upon atoms and our results in terms of a Boolean algebra. A quantum algorithm for evaluating simultaneously two of logical functions is proposed [55] Generalization of Deutsch's algorithm for determining all the mappings of a function is proposed [56]. Recently, the concrete quantum circuits for addition of any two numbers have been proposed [57]. In preparing all the augmentations for addition, we can expand any operation in mathematics. We expect its application for numerical computation. We lead to implementing commercial quantum-gated computers.

Some relations between Boolean algebra and quantum computing are discussed and proposed by Nagata and Nakamura. They show all the Boolean functions are set into the quantum computer just like the electronic computer. This fact means that all performances in logic of computing and control of itself are available even in quantum computers. Therefore, we could design any quantum-gated computer using the traditional design ways in the logic of existing electronic computers [53].

Further, they prove that the quantum computer can operate just like the electronic computer fundamentally through the operation of addition of two n-digit numbers. Therefore, the quantum computer can solve all the four basic operations of arithmetic, addition, subtraction, multiplication, and division. Further, it can be said that this quantum computer naturally operates not only arithmetic but also logic in terms of Boolean logic [57]. As a result, the theory proposed by [53, 57] can build a very true quantum-gated computer that is driven and operated by all software (all programs) used on existing electronic computers. Here, we point out how the argumentations are efficient when we compare our quantum computer with a classical method.

For efficiency of operations in a quantum oracle computing to the classical, we must note how many functions are used in an algorithm. Here, we sometimes see that the efficiency of this quantum operation seems to be N times the operation speed in classical operations, where N is the number of unknown factors. However, thinking of the Bernstein–Vazirani algorithm-like quantum computing, we could see the efficiency appears to be 2^N times the operation speed of the classical. In short, the efficiency depends upon the comparative object whenever we compare the two, which fact looks to be in our evaluations on comparison general.

We study an efficiency for operating a full adder/half adder by quantum-gated computing. A specific example is more understandable than the abstract structure when we study quantum computing theory. We investigate the concrete and specific example of the argumentations in reference [57]. Surprisingly, the concrete and specific calculation is faster than that of a classical apparatus which would require at least $2^{12} = 4096$ evaluations when we introduce the full adder operation [54]. Another concrete and specific calculation is faster than that of a classical apparatus which would require at least $2^8 = 256$ evaluations when we introduce only the half adder operation [54]. The quantum advantage increases when two numbers we treat become very large.

A quantum algorithm for a FULL adder operation based on registers of the CPU in a quantum-gated computer is discussed [58]. So far, the world of quantum-gated computing has been exploited, aimed at the limited applications based upon the Deutsch algorithm in 1985 and a few algorithms after his. However, the most important is to manage quantum-gated computers like we have designed and operated electronic computers. Fortunately, some theoretically novel challenge has been shown by Nagata and Nakamura. It happens that this challenge is too theoretical to be understood, but is very much needed in storing logical functions in Boolean algebra for the purpose of constructing hardware of quantum-gated computers and designing the programs. With storing data, the memory function is the basis of constructing quantum-gated computers. Toward practically simple construction of the quantum-gated computer, an example of simple memory, called registers, whose physical access is much easier is shown using the Bernstein–Vazirani algorithm.

Today computers are involved everywhere in life, where computers are electronic computers whose speed, however, is limited almost in most applications as we require higher and higher speed to satisfy our endless human desire. Today's computers are defined and successfully implemented, based on EDVAC (Electronic Discrete Variable Automatic Computer) by John von Neumann who focused on 'stored program concept' that had initially been proposed mathematically by Alan Turing, but is not practical in terms of implementation. However, von Neumann established a stored program computer and implemented it as EDVAC.

Meanwhile, it happens that these electronic computers are going to be replaced partially with quantum computers in some special applications like combinatorial problems. Today, we have two ways to develop quantum computers, one is by 'annealing' and the other is 'quantum-gated'. The former has fairly been developed and commercialized. But the latter is still in the development stage by IBM, Google, etc.

In 1985, Deutsch established the basis of a quantum algorithm for finding out the property of a function $f(x)$ within a quantum oracle computer regarded as 'quantum-gated' and not 'annealed'. This algorithm is fundamentally based on only 'XOR (exclusive OR)' to operate the application above, handling the two states $|0\rangle$ or $|1\rangle$ of one qubit instead of the two signals '0' or '1' in electronic computers. After that, a few modified quantum algorithms had appeared to increase the functionalities, but the applications are still limited.

Even in quantum computers like electronic computers, 'storing of programs and data' is the most important factor for data processing. Fortunately, we have discovered [53, 57] that the technique of using XOR based on Deutsch's algorithm can be expandable to storing any Boolean function, of course, including XOR in 'quantum memory' that automatically exists if we make use of the superposition and phase kickback concepts based upon XOR skillfully. It is important that such Boolean functions are regarded as 'basics of programs' for a general purpose computing even on quantum-gated computers.

It is quite natural that we can step out into the wider world as researchers because of having such memory in even quantum-gated computers. By virtue of having such memory, we can draw any quantum logic circuits of ALU (Arithmetic and Logic Unit) and the control unit, as the hardware of the CPU (Central Processing Unit) of the quantum-gated computer. And then on such a whole hardware as a computer, we may run any program to be designed and constructed using the logical concept of Boolean functions. When we have efficient coupling between the hardware and programs with data, we will be satisfied with attaining a quantum-gated computer.

Fortunately, we have discovered an algorithm for storing the logical functions in a Boolean algebra. In other words, this corresponds to discovery of memory for logic functions in quantum-gated computers by means of multiple operations based on exclusive OR with the two concepts, superposition and phase kickback. This theory is shown in reference [53].

In short, it is shown that all the Boolean functions are set into the quantum computer just like the electronic computer. This fact means that all performances in logic of computing and control of itself are available even in quantum computers. Therefore, we could design any quantum-gated computer using the traditional design ways in logic of existing electronic computers [53].

Based upon the above, we reviewed concretely how to store logical functions in the memory of the quantum-gated computer, thinking of not only logic circuits but also programs in terms of logic. And naturally we demonstrated typical arithmetic calculations in the binary system. From this trial, we are going to lead a quantum-gated computer to be implemented commercially. This is theoretically and practically shown in reference [57].

In more detail, we prove that the quantum computer can operate just like the electronic computer fundamentally through the operation of addition of two *n*-digit numbers. Therefore, the quantum computer can solve all the four basic operations of arithmetic, addition, subtraction, multiplication, and division. Further, it can be said that this quantum computer naturally operates not only arithmetic but also logic in terms of Boolean logic [57].

As a result, the theory proposed by references [53, 57] can build a quantum-gated computer that is driven and operated by all software (all programs) used on existing electronic computers.

Finally, we say that such quantum-gated computers are much faster than existing electronic computers by virtue of quantum effectiveness, that quantum mechanics consists of the most important concept of an elementary particle much faster than the speed of light. So, we could expand processing functionality of quantum-gated computers for a wider range of applications based on physics and chemistry. In addition, we might guess that the energy consumption of quantum computers would be much smaller than that of electronic computers.

The next step is experimental demonstrations of our quantum-gated computer. We have said that all the Boolean functions are set into the quantum computer just like the electronic computer by using the quantum phase factor [53]. But, in a real experiment, it is somewhat difficult to evaluate the value of the quantum phase factor. Therefore, we show all the Boolean functions can be set into the quantum computer by using the quantum superposition instead of the quantum phase factor. This is the key for connecting our theory into real experimental demonstrations. Further, we can demonstrate typical arithmetic calculations [57] using the superposition method as well.

The advent of quantum information theory [59–62] leads to exciting new developments that are either superior to classical approaches or have no classical counterpart, such as quantum cryptographic key distribution [63], quantum teleportation [64], superdense coding [65], and quantum computation [66, 67]. In particular, in the last of these cases, the combination of quantum correlation and the quantum superposition principle challenges the very concept of an algorithm as a sequence of deterministic orders à la Turing. This is because the underlying hypotheses of algorithms conventionally understood rely implicitly on the laws of classical physics. In fact, Feynman [68] first suggested that computational devices, behaving according to quantum mechanics, could be qualitatively more powerful than classical computers.

In 1993, Bernstein and Vazirani [11, 12] were among the first to show (although David Deutsch [8] introduced the Deutsch–Turing thesis earlier in 1985) that a quantum Turing machine would be capable of simulating quantum mechanical systems in polynomial time, which constitutes an exponential improvement beyond classical approaches. Later, quantum algorithms developed by Shor [15] and Grover [16] provided a strong motivation for exploring the computational possibilities offered by quantum mechanics. The crucial ingredients in these algorithms, namely, amplitude amplification in the case of quantum search and phase estimation in the case of quantum factorization, are important elements that are common to many current quantum algorithms [6, 69, 70].

Many other quantum algorithms have been developed since Feynman's insight, notably the Deutsch–Jozsa algorithm [9, 10], its variations, and extensions. For instance, the Deutsch–Jozsa algorithm can be used for quantum key distribution [71] and for a secure quantum key distribution based on the use of an entangled state [72]. Additionally, a method of a secure quantum cryptography based on the Deutsch–Jozsa algorithm is proposed by Nagata *et al* [73]. The ultimate goal of quantum algorithms is of providing results superior to those achievable by their classical counterparts, preferably with an exponential advantage.

Quantum communication is the art of transferring a quantum state from one place to another. Traditionally, the sender is named Alice and the receiver Bob. The basic motivation is that quantum states code quantum information—called qubits in the case of two-dimensional Hilbert spaces—and that quantum information allows one to perform tasks that could only be achieved far less efficiently, if at all, using classical information. In 2015, Nagata and Nakamura [71] discussed the use of the Deutsch–Jozsa algorithm for quantum key distribution, and in 2017, they described a method of secure quantum key distribution based on Deutsch's algorithm using an entangled state [72]. Subsequently, Nagata *et al* [73, 74] proposed an approach to high-speed secure quantum cryptography based on the Deutsch–Jozsa algorithm. And a generalization of it in *d*-level quantum system is explored by Nguyen and Kim [75, 76]. Quantum communication based on an algorithm of determining a matrix was proposed [77]. Quantum communication based on Simon's algorithm was studied [78]. The relation between quantum computers and secret sharing based on the use of quantum principles was discussed by Diep and Giang [79]. An application of quantum Gauss–Jordan elimination code to quantum secret sharing code was studied [80]. Quantum cryptography by means of quantum computer algorithms was proposed by Nagata *et al* [81] in 2020. Quantum cryptography based on an algorithm for determining a function using qudit systems was studied by Nagata *et al* [82].

Continuous-variable quantum information is the area of quantum information science that makes use of physical observables, such as the strength of an electromagnetic field, whose numerical values belong to continuous intervals. In 1998, Braunstein studied error correction for continuous quantum variables [83] and quantum error correction for communication with linear optics [84]. In 1999, Lloyd and Braunstein proposed quantum computation over continuous variables [85]. The same year, Ralph considered continuous-variable quantum cryptography [86]. In 2000, Hillery discussed quantum cryptography with squeezed states [87], while Reid described quantum cryptography with a predetermined key using continuous-variable Einstein–Podolsky–Rosen correlations [88]. In 2001, secure quantum key distribution using squeezed states was studied by Gottesman and Preskill [89].

A year later, continuous-variable quantum cryptography using coherent states was first proposed by Grosshans and Grangier [90]. Efficient classical simulation of continuous-variable quantum information processes has been studied by Bartlett *et al* [91]. Continuous-variable quantum computing and its applications to cryptography were proposed by Diep *et al* [92].

Measurement theory in Deutsch's algorithm based on the truth values was discussed by Nagata and Nakamura [93]. The projective measurement theory is indeed a successful quantum measurement theory. The projective measurement theory gives approximate and at times remarkably accurate numerical predictions. Much experimental data approximately fits to the projective measurement theory for the past some 100 years. We do not doubt the correctness of the project measurement theory. It is one of the consistent quantum measurement theories.

We propose a new measurement theory, in qubits handling, based on the truth values, i.e., the truth T (1) for true and the falsity F (0) for false. The results of measurement are either 0 or 1. To implement Deutsch's algorithm, we need both observability and controllability of a quantum state. The new measurement theory can satisfy these two. Especially, we systematically describe our assertion based on more mathematical analysis using raw data in a thoughtful experiment.

In this work, we present a method for secure quantum key distribution based on a special Deutsch–Jozsa algorithm using Greenberger–Horne–Zeilinger states [94, 95]. The security of the method relies on the Ekert 91 protocol [63]: i.e., an eavesdropper must destroy the GHZ state. We discuss how the Bernstein–Vazirani algorithm can be used for quantum communication with the inclusion of error correction, and we study this algorithm in a noisy environment. We propose a method for computing many functions using the generalized Bernstein–Vazirani algorithm and many parallel quantum systems, and we use this method to perform multiplication operations. We study quantum communication based on Simon's algorithm and generalize Deutsch's algorithm for determining all the mappings of a function. Proposing a parallel computation for all of the combinations of values in variables of a logical function, we demonstrate some parallel evaluations of two two-variable functions which are elements of a Boolean algebra composed of the four-atom set. Using quantum cryptography based on an algorithm for determining a function using qudit systems, we propose continuous-variable quantum computing and its applications to cryptography. We present various new forms of the Bernstein–Vazirani algorithm beyond qubit systems. We propose a number of new quantum algorithms: one for rapidly determining a homogeneous linear function, one for finding the roots of a polynomial function, and one for rapidly plotting a function. Considering a quantum algorithm for determining whether or not a given function is even, we propose a necessary and sufficient condition for quantum computing. Finally, studying our problem toward practical quantum-gated computers, we propose the concrete quantum circuits for addition of any two numbers and we study the computational complexity of this. Additionally, we study the measurement theory based on the truth values.

Parts of this chapter have been reprinted from [58].

References

[1] von Neumann J 1955 *Mathematical Foundations of Quantum Mechanics* (Princeton, NJ: Princeton University Press)

[2] Feynman R P, Leighton R B and Sands M 1965 *Lectures on Physics* Volume III (Reading, MA: Addison-Wesley)

[3] Redhead M 1989 *Incompleteness, Nonlocality, and Realism* 2nd edn (Oxford: Clarendon)

[4] Peres A 1993 *Quantum Theory: Concepts and Methods* (Dordrecht: Kluwer)

[5] Sakurai J J 1995 *Modern Quantum Mechanics* revised edn (Reading, MA: Addison-Wesley)

[6] Nielsen M A and Chuang I L 2000 *Quantum Computation and Quantum Information* (Cambridge: Cambridge University Press)

[7] Rennie R (ed) 2015 *Oxford Dictionary of Physics* 7th edn (Oxford: Oxford University Press)

[8] Deutsch D 1985 Quantum theory, the Church–Turing principle and the universal quantum computer *Proc. R. Soc.* A **400** 97

[9] Deutsch D and Jozsa R 1992 Rapid solution of problems by quantum computation *Proc. R. Soc. A* 439 553

[10] Cleve R, Ekert A, Macchiavello C and Mosca M 1998 Quantum algorithms revisited *Proc. R. Soc.* A **454** 339

[11] Bernstein E and Vazirani U 1993 Quantum complexity theory *Proc. of 25th Annual ACM Symp. on Theory of Computing (STOC '93)* 11

[12] Bernstein E and Vazirani U 1997 Quantum complexity theory *SIAM J. Comput.* **26** 1411

[13] Nagata K, Resconi G, Nakamura T, Batle J and Abdalla S 2017 A generalization of the Bernstein–Vazirani algorithm *MOJ Ecol. Environ. Sci.* **2** 00010

[14] Simon D R 1994 On the power of quantum computation *Proc. of 35th IEEE Annual Symp. on Foundations of Computer Science* 116

[15] Shor P W 1994 Algorithms for quantum computation: discrete logarithms and factoring *Proc. of 35th IEEE Annual Symp. on Foundations of Computer Science* 124

[16] Grover L K 1996 A fast quantum mechanical algorithm for database search *Proc. of 28th Annual ACM Symp. on Theory of Computing* 212

[17] Jones J A and Mosca M 1998 Implementation of a quantum algorithm on a nuclear magnetic resonance quantum computer *J. Chem. Phys.* **109** 1648

[18] Gulde S, Riebe M, Lancaster G P T, Becher C, Eschner J, Häffner H, Schmidt-Kaler F, Chuang I L and Blatt R 2003 Implementation of the Deutsch–Jozsa algorithm on an ion-trap quantum computer *Nature* **421** 48

[19] de Oliveira A N, Walborn S P and Monken C H 2005 Implementing the Deutsch algorithm with polarization and transverse spatial modes *J. Opt. B: Quantum Semiclass. Opt* **7** 288

[20] Kim Y-H 2003 Single-photon two-qubit entangled states: preparation and measurement *Phys. Rev.* A **67** 040301

[21] Mohseni M, Lundeen J S, Resch K J and Steinberg A M 2003 Experimental application of decoherence-free subspaces in an optical quantum-computing algorithm *Phys. Rev. Lett.* **91** 187903

[22] Tame M S, Prevedel R, Paternostro M, Böhi P, Kim M S and Zeilinger A 2007 Experimental realization of Deutsch's algorithm in a one-way quantum computer *Phys. Rev. Lett.* **98** 140501

[23] Du J, Shi M, Zhou X, Fan Y, Ye B J, Han R and Wu J 2001 Implementation of a quantum algorithm to solve the Bernstein-Vazirani parity problem without entanglement on an ensemble quantum computer *Phys. Rev.* A **64** 042306

[24] Brainis E, Lamoureux L-P, Cerf N J, Emplit P H, Haelterman M and Massar S 2003 *Phys. Rev. Lett.* **90** 157902

[25] Londero P, Dorrer C, Anderson M, Wallentowitz S, Banaszek K and Walmsley I A 2004 Efficient optical implementation of the Bernstein–Vazirani algorithm *Phys. Rev.* A **69** 010302

[26] Cross A W, Smith G and Smolin J A 2015 Quantum learning robust against noise *Phys. Rev.* A **92** 012327

[27] Li H and Yang L 2015 A quantum algorithm for approximating the influences of Boolean functions and its applications *Quantum Inf. Process.* **14** 1787

[28] Adcock M R A, Hoyer P and Sanders B C 2016 Quantum computation with coherent spin states and the close Hadamard problem *Quantum Inf. Process.* **15** 1361

[29] Fallek S D, Herold C D, McMahon B J, Maller K M, Brown K R and Amini J M 2016 Transport implementation of the Bernstein-Vazirani algorithm with ion qubits *New J. Phys.* **18** 083030

[30] Krishna R, Makwana V and Suresh A P 2016 A generalization of Bernstein-Vazirani algorithm to qudit systems arXiv:1609.03185

[31] Mehendale D P and Joag P S 2017 A simple algorithm for complete factorization of an N-partite pure quantum state *Quantum Phys. Lett.* **6** 73

[32] Fujikawa K, Oh C H and Umetsu K 2019 A classical limit of Grover's algorithm induced by dephasing: coherence versus entanglement *Mod. Phys. Lett.* A **34** 1950146

[33] Yin A, He K and Fan P 2019 Quantum dialogue protocol based on Grover's search algorithms *Mod. Phys. Lett.* A **34** 1950169

[34] Diep D N, Giang D H and Van Minh N 2017 Quantum Gauss-Jordan elimination and simulation of accounting principles on quantum computers *Int. J. Theor. Phys.* **56** 1948

[35] Jin W 2016 Dynamical analysis of Grover's search algorithm in arbitrarily high-dimensional search spaces *Quantum Inf. Process.* **15** 65

[36] Nagata K, Resconi G, Nakamura T, Batle J, Abdalla S, Farouk A and Geurdes H 2017 A method of computing many functions simultaneously by using many parallel quantum systems Asian *J. Math. Phys.* **1** 1 https://mathphys.asia/index.php/volume-1-2017/

[37] Yasudo R, Matsutani H, Koibuchi M, Amano H and Nakamura T 2017 Scalable networks-on-chip with elastic links demarcated by decentralized routers *IEEE Trans. Comput.* **66** 702–16

[38] De Broglie-Bohm Theory Wikipedia. The Free Encyclopedia https://en.wikipedia.org/wiki/De-Broglie-Bohm-theory

[39] Schon C and Beige A 2001 Analysis of a two-atom double-slit experiment based on environment-induced measurements *Phys. Rev.* A **64** 023806

[40] Nagata K, Nakamura T, Geurdes H, Batle J, Abdalla S, Farouk A and Diep D N 2018 Creating very true quantum algorithms for quantum energy based computing *Int. J. Theor. Phys.* **57** 973

[41] Nagata K, Nakamura T, Geurdes H, Batle J, Abdalla S and Farouk A 2018 New method of calculating a multiplication by using the generalized Bernstein-Vazirani algorithm *Int. J. Theor. Phys.* **57** 1605

[42] Nagata K and Nakamura T 2018 A new mathematical structure for quantum algorithms in case of a special function *J. Sci. Eng. Res.* **5** 326–8 http://jsaer.com/download/vol-5-iss-3-2018/JSAER2018-05-03-326-328.pdf

[43] Nagata K and Nakamura T 2018 Standard quantum algorithms and the property of a certain function *Phys. Astron. Int. J.* **2** 146–7

[44] Nagata K, Nakamura T, Batle J and Farouk A 2018 Efficient quantum algorithm for the parity problem of a certain function *Int. J. Theor. Phys.* **57** 3098
[45] Nagata K, Nakamura T, Farouk A and Diep D N 2019 Necessary and sufficient condition for quantum computing *Int. J. Theor. Phys.* **58** 136
[46] Devitt S J, Munro W J and Nemoto K 2013 Quantum error correction for beginners *Rep. Prog. Phys.* **76** 076001
[47] Nagata K, Nakamura T, Geurdes H, Batle J, Farouk A, Diep D N and Patro S K 2018 Efficient quantum algorithms of finding the roots of a polynomial function *Int. J. Theor. Phys.* **57** 2546
[48] Nagata K and Nakamura T 2019 Quantum algorithm for the root-finding problem *Quantum Stud.: Math. Found.* **6** 135
[49] Resconi G and Nagata K 2018 Quantum circuit by one step method and similarity with neural network *Int. J. Gen. Eng. Technol* **7** 1–20 https://www.iaset.us/archives?jname=74−2&year=2018&submit=Search)
[50] Nagata K, Geurdes H, Patro S K, Heidari S, Farouk A and Nakamura T 2019 Quantum algorithm for determining a complex number string *Int. J. Theor. Phys.* **58** 3694
[51] Nagata K, Geurdes H, Patro S K, Heidari S, Farouk A and Nakamura T 2020 Generalization of the Bernstein-Vazirani algorithm beyond qubit systems *Quantum Stud.: Math. Found.* **7** 17
[52] Nagata K, Patro S K, Geurdes H, Heidari S, Diep D N and Nakamura T 2019 Various new forms of the Bernstein-Vazirani algorithm beyond qubit systems Asian *J. Math. Phys.* **3** 1 https://mathphys.asia/index.php/volume-3-2019/
[53] Nagata K and Nakamura T 2020 Some theoretically organized algorithm for quantum computers *Int. J. Theor. Phys.* **59** 611
[54] Gilbert W J and Nicholson W K 2004 *Modern Algebra with Applications* 2nd edn (New York: Wiley)
[55] Nagata K and Nakamura T 2020 Quantum algorithm for evaluating two of logical functions simultaneously *Int. J. Theor. Phys.* **59** 2191
[56] Nagata K and Nakamura T 2020 Generalization of Deutsch's algorithm *Int. J. Theor. Phys.* **59** 2557
[57] Nakamura T and Nagata K 2021 Physics' evolution toward computing *Int. J. Theor. Phys.* **60** 70
[58] Nagata K and Nakamura T 2021 A quantum algorithm for a FULL adder operation based on registers of the CPU in a quantum-gated computer *Int. J. Theor. Phys.* **60** 2986–94
[59] Lo H-K, Popescu S and Spiller T (ed) 1998 *Introduction to Quantum Computation and Information* (Singapore: World Scientific)
[60] Galindo A and Martín-Delgado M A 2002 Information and computation: classical and quantum aspects *Rev. Mod. Phys.* **74** 347
[61] Williams C P and Clearwater S H 1997 *Explorations in Quantum Computing* (New York: Springer)
[62] Williams C P (ed) 1998 *Quantum Computing and Quantum Communications* (Berlin: Springer)
[63] Ekert A K 1991 Quantum cryptography based on Bell's theorem *Phys. Rev. Lett.* **67** 661
[64] Bennett C H, Brassard W K, Crepeau G, Jozsa C, Peres R and Wootters A 1993 Teleporting an unknown quantum state via dual classical and Einstein-Podolsky-Rosen channels *Phys. Rev. Lett.* **70** 1895

[65] Bennett C H and Wiesner S J 1993 Communication via one- and two-particle operators on Einstein-Podolsky-Rosen states *Phys. Rev. Lett.* **69** 2881
[66] Ekert A and Jozsa R 1996 Quantum computation and Shor's factoring algorithm *Rev. Mod. Phys.* **68** 733
[67] Berman G P, Doolen G D, Mainieri R and Tsifrinovich V I 1998 *Introduction to Quantum Computers* (Singapore: World Scientific)
[68] Feynman R 1982 Simulating physics with computers *Int. J. Theor. Phys.* **21** 467
[69] Smith J and Mosca M 2012 Algorithms for quantum computers *Handbook of Natural Computing* (Cham: Springer) p 1451
[70] Childs A M and Van Dam W 2010 Quantum algorithms for algebraic problems *Rev. Mod. Phys.* **82** 1
[71] Nagata K and Nakamura T 2015 The Deutsch–Jozsa algorithm can be used for quantum key distribution *Open Access Library J.* **2** e1798
[72] Nagata K and Nakamura T 2017 Quantum cryptography, quantum communication, and quantum computer in a noisy environment *Int. J. Theor. Phys.* **56** 2086
[73] Nagata K, Nakamura T and Farouk A 2017 Quantum cryptography based on the Deutsch–Jozsa algorithm *Int. J. Theor. Phys.* **56** 2887
[74] Nagata K, Nakamura T, Geurdes H, Batle J, Abdalla S and Farouk A 2018 Secure quantum key distribution based on a special Deutsch–Jozsa algorithm *Asian J. Math. Phys.* **2** 6 https://mathphys.asia/index.php/volume-2-2018/
[75] Nguyen D M and Kim S 2019 Quantum key distribution protocol based on modified generalization of Deutsch-Jozsa algorithm in *d*-level quantum system *Int. J. Theor. Phys.* **58** 71
[76] Nguyen D M and Kim S 2019 Multi-bits transfer based on the quantum three-stage protocol with quantum error correction codes *Int. J. Theor. Phys.* **58** 2043
[77] Nagata K and Nakamura T 2019 Quantum communication based on an algorithm of determining a matrix *Int. J. Theor. Phys.* **58** 247
[78] Nagata K, Nakamura T, Geurdes H, Batle J, Abdalla S and Farouk A 2017 Quantum communication based on Simon's algorithm *Int. J. Emerg. Eng. Res. Technol* **5** 28–31 https://www.ijeert.org/v5-i8
[79] Diep D N and Giang D H 2017 Quantum communication and quantum multivariate polynomial interpolation *Int. J. Theor. Phys.* **56** 2797
[80] Diep D N, Giang D H and Phu P H 2018 Application of quantum Gauss-Jordan elimination code to quantum secret sharing code *Int. J. Theor. Phys.* **57** 841
[81] Nagata K, Diep D N and Nakamura T 2020 Quantum cryptography through some algorithm for quantum computers *Asian J. Math. Phys.* **4** 7 https://mathphys.asia/index.php/volume-4-2020/
[82] Nagata K, Diep D N and Nakamura T 2020 Quantum cryptography based on an algorithm for determining a function using qudit systems *Int. J. Theor. Phys.* **59** 2875
[83] Braunstein S L 1998 Error correction for continuous quantum variables *Phys. Rev. Lett.* **80** 4084
[84] Braunstein S L 1998 Quantum error correction for communication with linear optics *Nature (London)* **394** 47
[85] Lloyd S and Braunstein S L 1999 Quantum computation over continuous variables *Phys. Rev. Lett.* **82** 1784
[86] Ralph T C 1999 Continuous variable quantum cryptography *Phys. Rev.* A **61** 010303

[87] Hillery M 2000 Quantum cryptography with squeezed states *Phys. Rev.* A **61** 022309

[88] Reid M D 2000 Quantum cryptography with a predetermined key, using continuous-variable Einstein-Podolsky-Rosen correlations *Phys. Rev.* A **62** 062308

[89] Gottesman D and Preskill J 2001 Secure quantum key distribution using squeezed states *Phys. Rev.* A **63** 022309

[90] Grosshans F and Grangier P 2002 Continuous variable quantum cryptography using coherent states *Phys. Rev. Lett.* **88** 057902

[91] Bartlett S D, Sanders B C, Braunstein S L and Nemoto K 2002 Efficient classical simulation of continuous variable quantum information processes *Phys. Rev. Lett.* **88** 097904

[92] Diep D N, Nagata K and Wong R 2020 Continuous-variable quantum computing and its applications to cryptography *Int. J. Theor. Phys.* **59** 3184

[93] Nagata K and Nakamura T 2016 Measurement theory in Deutsch's algorithm based on the truth values *Int. J. Theor. Phys.* **55** 3616

[94] Greenberger D M, Horne M A and Zeilinger A 1989 *Bell's Theorem, Quantum Theory and Conceptions of the Universe* ed M Kafatos (Dordrecht: Kluwer) p 69

[95] Greenberger D M, Horne M A, Shimony A and Zeilinger A 1990 Bell's theorem without inequalities *Am. J. Phys.* **58** 1131

IOP Publishing

Simplified Quantum Computing with Applications

Koji Nagata, Do Ngoc Diep, Ahmed Farouk and Tadao Nakamura

Chapter 2

Overview figures for a method of understanding quantum computing

This chapter provides representative overview figures for readers' convenience in understanding especially two basic quantum algorithms strongly related to all the concepts throughout this book.

2.1 What quantum-gated computing needs in its algorithms

Superposition and phase kickback are most needed to operate quantum states of the quantum gate at any stage of a quantum circuit. There are two types of qubits in a two-qubit quantum gate, where one is a target qubit (sometimes, called a controlled qubit), while the other is a control qubit.

- Superposition is a natural concept as long as one is handling quantum computation (operations). In particular, the Hadamard gate contributes to superposition in order to make very true quantum operations with probabilistic behaviors.
- Phase kickback is a very convenient concept between a control qubit and a target qubit, which fact makes good use of a quantum effect concerning the eigenstate and the eigenvalue.

2.2 Every reversibility in quantum circuits is by virtue of exclusive OR

It is exclusive OR $\oplus$ that makes reversibility in unitary operations U_f reversible (figure 2.1). It is possible to transform the state $|x, y\rangle$ into

$$|x, y \oplus f(x)\rangle, \tag{2.2.1}$$

by applying U_f. We denote the transformation U_f defined by the map

$$U_f|x, y\rangle = |x, y \oplus f(x)\rangle. \tag{2.2.2}$$

doi:10.1088/978-0-7503-4700-6ch2

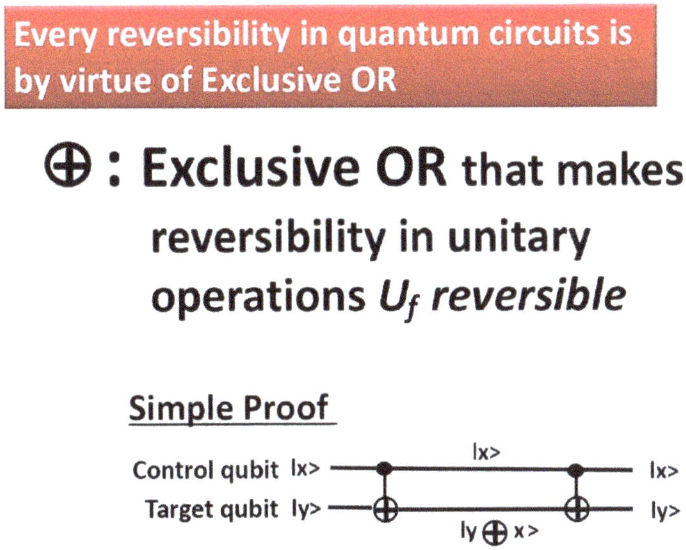

Figure 2.1. Every reversibility in quantum circuits is by virtue of exclusive OR.

Thus, we have

$$(U_f)^2|x, y\rangle = |x, y \oplus 2f(x)\rangle = |x, y\rangle. \tag{2.2.3}$$

Therefore, $(U_f)^2 = I$ and we can say unitary operations U_f are reversible.

2.3 Equivalence of the circuits by virtue of superposition of qubits to be applied by Hadamard gates

In this section, we study Deutsch's algorithm and the Bernstein–Vazirani algorithm together in terms of deep understanding of quantum computing. They use Hadamard gates. We cannot perform them without Hadamard gates because we must treat phase kickback to operate and solve some problems using Deutsch's algorithm. Thus, Hadamard gates are very important for understanding these two basic quantum algorithms. This equivalence is for understanding how the relation between the control qubit and the target qubit becomes interchanged under the phase kickback, and then the result of the operation of either algorithm is on the first bit of the output (figure 2.2).

2.4 Bases of quantum computing

Here we have a good chance to consider the origin, or rather the motive of Deutsch's algorithm. The algorithm seems to be made based on CNOT that is consisting of the control qubit $|x\rangle$ and controlled qubit $|y\rangle$ changing into $|y \oplus x\rangle$. The function from its input to output is easily implemented using ordinary electronic devices, and also the function is shown using a matrix even in a quantum operation way. However, this operation is very limited to usual classical function. Meanwhile, the presentation

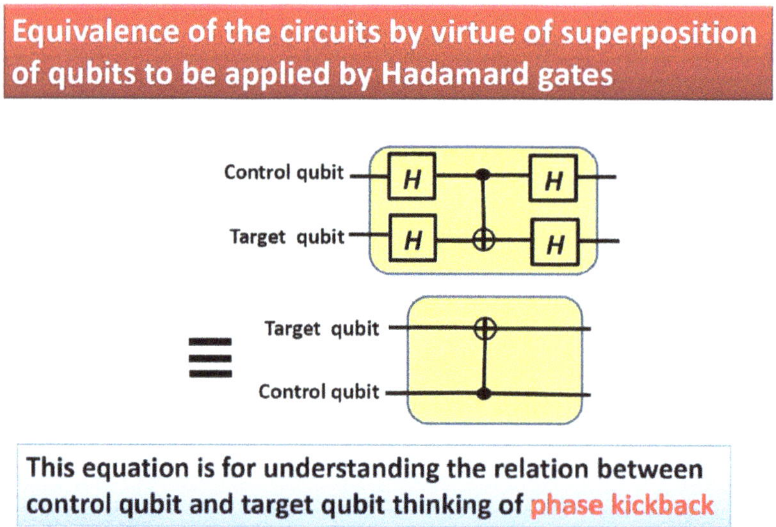

Figure 2.2. Equivalence of the circuits by virtue of superposition of qubits to be applied by Hadamard gates.

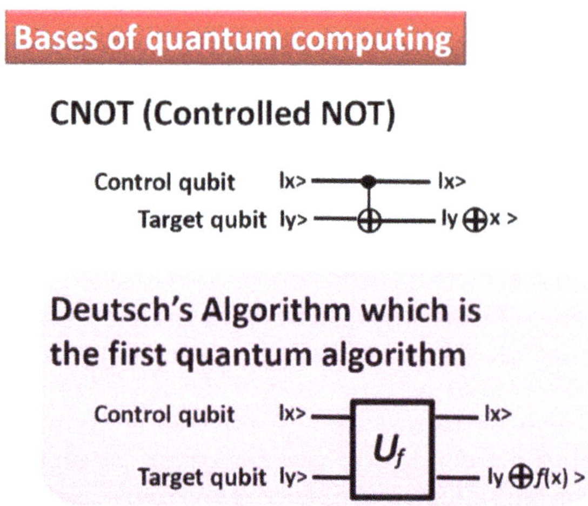

Figure 2.3. Bases of quantum computing.

power of this Deutsch's model seems to be attractive in quantum operations as a much higher level (figure 2.3).

We here do imagine the motive of Deutsch's algorithm could be based on CNOT, but the function of it should be more expandable to handle the algorithm more skillfully in terms of a couple of control and controlled target qubits in quantum operations. Instead of the matrix as a likely quantum oracle in CNOT, he introduced a more expandable quantum oracle U_f which can describe more detailed actions based on quantum mechanics behavior using the concepts eigenstate and

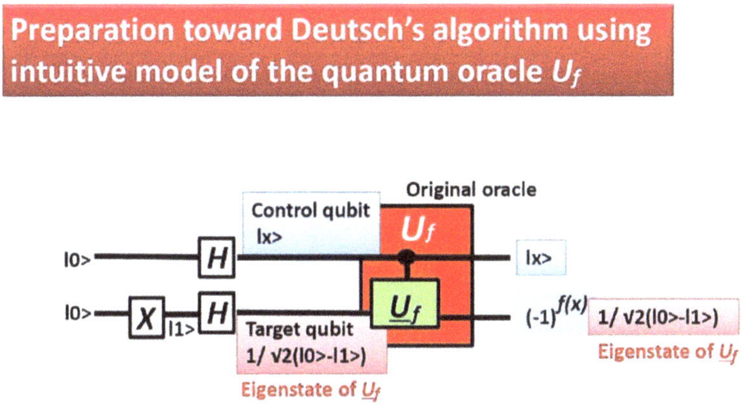

Figure 2.4. Preparation toward Deutsch's algorithm using intuitive model of the quantum oracle U_f. Reproduced with permission from [1].

eigenvalue, which are related to keeping the second qubit of the oracle the same and phase kickback, respectively. To have this action the concept of superposition is necessary everywhere in the circuit. As a result, Deutsch expanded the function of CNOT to the oracle U_f to be able to control the target (controlled) qubit by $f(x)$ and not a single x in mapping $|x\rangle|y\rangle$ to $|x\rangle|y \oplus f(x)\rangle$.

2.5 Preparation toward Deutsch's algorithm using intuitive model of the quantum oracle U_f

The interpretation of oracle U_f is not always clear because the first qubit of the oracle is the same at the right side (output side) of this, even though the same state name $|x\rangle$ is used at both without the straight direct connection in the oracle U_f (figure 2.4).

In order to clear the function as the operation, the intuitive model is introduced. So, this figure says that, so long as it is using $1/\sqrt{2}(|0\rangle - |1\rangle)$ as its eigenstate for the intuitive oracle gate, $1/\sqrt{2}(|0\rangle - |1\rangle)$ is changed into the $(-1)^{f(x)}1/\sqrt{2}(|0\rangle - |1\rangle)$ that has the eigenvalue factor $(-1)^{f(x)}$ after operation by the intuitive oracle U_f gate. However, the eigenstate should not be changed because of the proper equation based on quantum mechanics. Thus, the eigenvalue factor $(-1)^{f(x)}$ goes to the first qubit and it is changed into the $(-1)^{f(x)}|x\rangle$. This is the phase kickback described in the next section [1].

2.6 Preparation with phase kickback toward Deutsch's algorithm using an intuitive model of the quantum oracle U_f

Generally speaking, so long as using $1/\sqrt{2}(|0\rangle - |1\rangle)$ as its eigenstate for a gate, $|x\rangle$ of first qubit is changed into the $(-1)^{f(x)}|x\rangle$ that has the phase kickback factor $(-1)^{f(x)}$ after operation by the gate. Thus, we can say that quantum computing is the proper equation based on quantum mechanics (figure 2.5).

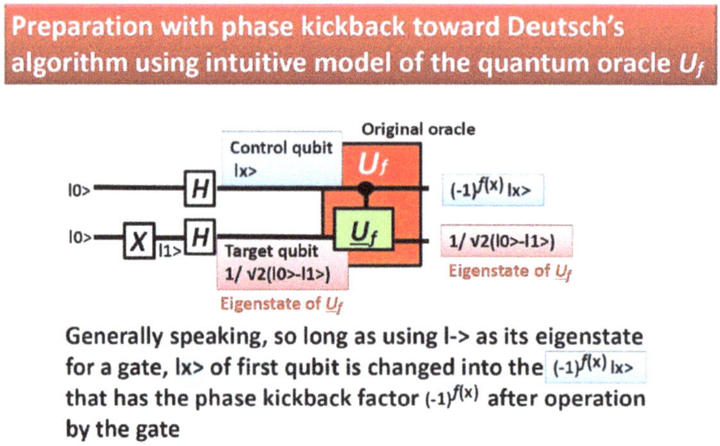

Figure 2.5. Preparation with phase kickback toward Deutsch's algorithm using intuitive model of the quantum oracle U_f. Reproduced with permission from [1].

2.7 Deutsch's algorithm

We notice

$$\sum_{x=0,1} (-1)^{f(x)}|x\rangle = \pm(|0\rangle + |1\rangle) \tag{2.7.1}$$

if $f(0) = f(1)$, that is, f is constant.

We notice

$$\sum_{x=0,1} (-1)^{f(x)}|x\rangle = \pm(|0\rangle - |1\rangle) \tag{2.7.2}$$

if $f(0) \neq f(1)$, that is, f is balanced.

Here the story of Deutsch's algorithm has arrived. Figure 2.6 shows the overview of Deutsch's algorithm finally including the measurement. Again, generally speaking, as long as it is using $1/\sqrt{2}(|0\rangle - |1\rangle)$ as its eigenstate for an oracle gate U_f, $|x\rangle$ of first qubit is changed into the $(-1)^{f(x)}|x\rangle$ that has the phase kickback factor $(-1)^{f(x)}$ after operation by the gate.

2.8 Bernstein–Vazirani algorithm—general expression by eigenstate concept

Let us consider three positions in the figure for the quantum states A, B, and C (figure 2.7). We have

- In state A is $|0\rangle^{\otimes 3}$.
- In state B is $|+\rangle^{\otimes 3}$.
- In state C is

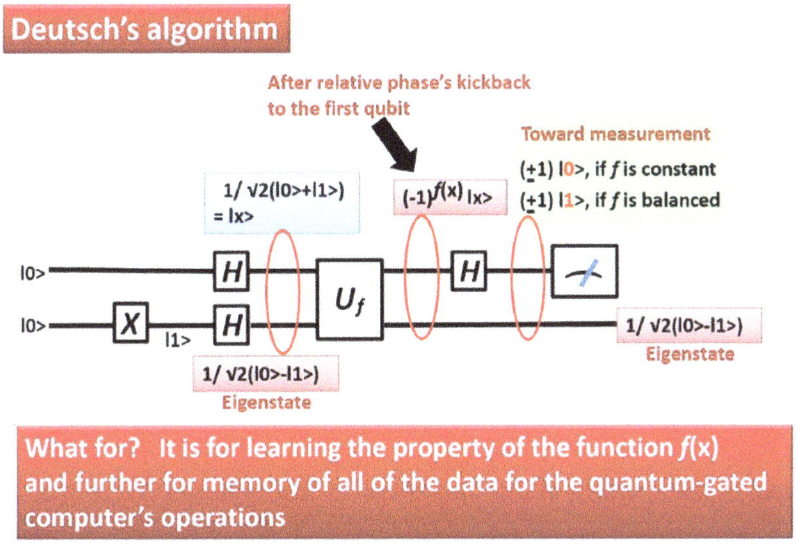

Figure 2.6. Deutsch's algorithm.

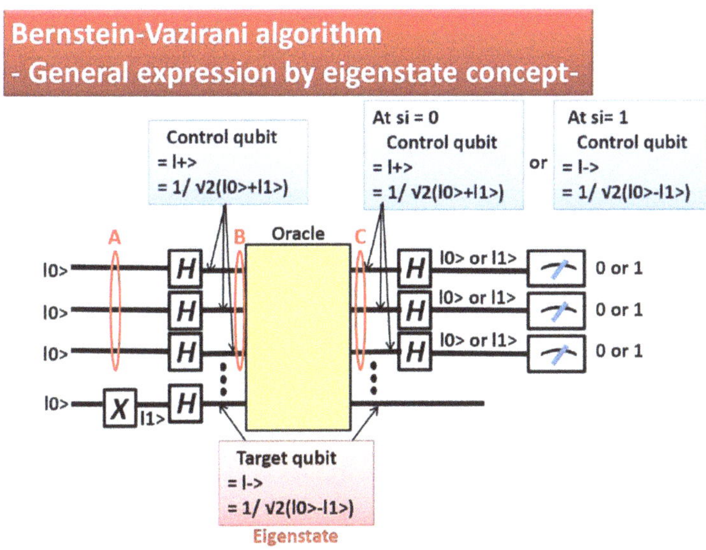

Figure 2.7. Bernstein–Vazirani algorithm—general expression by eigenstate concept.

$$\sum_{x\in\{0,1\}^3} \frac{(-1)^{f(x)}|x\rangle}{\sqrt{2^3}} = \sum_{x\in\{0,1\}^3} \frac{(-1)^{s\cdot x}|x\rangle}{\sqrt{2^3}}$$
$$= \frac{1}{\sqrt{2}}(|0\rangle + (-1)^{s_1}|1\rangle) \otimes \frac{1}{\sqrt{2}}(|0\rangle + (-1)^{s_2}|1\rangle) \quad (2.8.1)$$
$$\otimes \frac{1}{\sqrt{2}}(|0\rangle + (-1)^{s_3}|1\rangle).$$

Here, the secret string is $s = s_1s_2s_3$.

2.9 Implementation of the phase oracle based on CNOT for the Bernstein–Vazirani algorithm

Note the Bernstein–Vazirani algorithm is directly based on the CNOT gate (figure 2.8).

Control qubit is defined as follows:

$$|+\rangle \equiv 1/\sqrt{2}(|0\rangle + |1\rangle) \equiv 1/\sqrt{2}\begin{pmatrix}1\\1\end{pmatrix}. \quad (2.9.1)$$

Target qubit is defined as follows:

$$|-\rangle \equiv 1/\sqrt{2}(|0\rangle - |1\rangle) \equiv 1/\sqrt{2}\begin{pmatrix}1\\-1\end{pmatrix}. \quad (2.9.2)$$

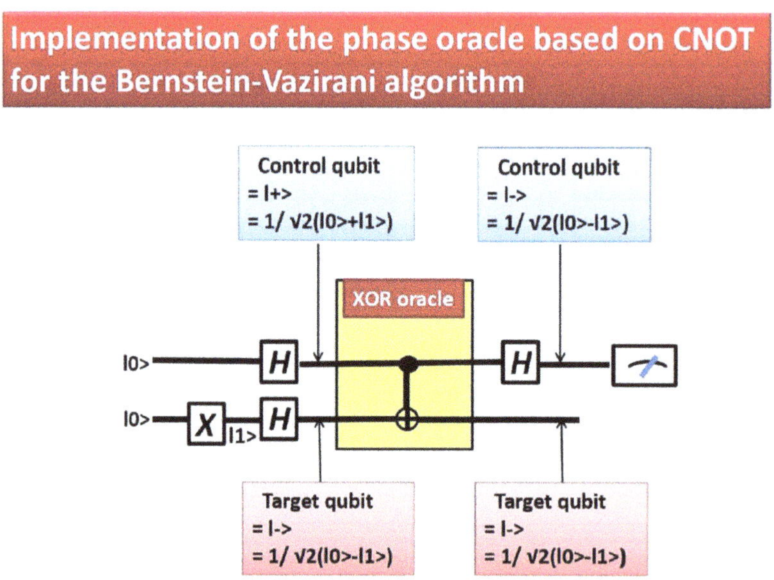

Figure 2.8. Implementation of the phase oracle based on CNOT.

Then, the two-qubit system has the following joint state:

$$|+\rangle| - \rangle = 1/\sqrt{2}(|0\rangle + |1\rangle) \otimes 1/\sqrt{2}(|0\rangle - |1\rangle)$$
$$= 1/2\begin{pmatrix} 1 \times \begin{pmatrix} 1 \\ -1 \end{pmatrix} \\ 1 \times \begin{pmatrix} 1 \\ -1 \end{pmatrix} \end{pmatrix} = 1/2\begin{pmatrix} 1 \\ -1 \\ 1 \\ -1 \end{pmatrix}. \tag{2.9.3}$$

The matrix form of CNOT is defined as follows:

$$\text{CNOT} \equiv \begin{pmatrix} 1 & 0 & 0 & 0 \\ 0 & 1 & 0 & 0 \\ 0 & 0 & 0 & 1 \\ 0 & 0 & 1 & 0 \end{pmatrix}. \tag{2.9.4}$$

After phase oracle CNOT, the joint state becomes as follows:

$$\begin{pmatrix} 1 & 0 & 0 & 0 \\ 0 & 1 & 0 & 0 \\ 0 & 0 & 0 & 1 \\ 0 & 0 & 1 & 0 \end{pmatrix} 1/2\begin{pmatrix} 1 \\ -1 \\ 1 \\ -1 \end{pmatrix} = 1/2\begin{pmatrix} 1 \\ -1 \\ -1 \\ 1 \end{pmatrix} = 1/2\begin{pmatrix} 1 \times \begin{pmatrix} 1 \\ -1 \end{pmatrix} \\ -1 \times \begin{pmatrix} 1 \\ -1 \end{pmatrix} \end{pmatrix} \tag{2.9.5}$$
$$= 1/\sqrt{2}(|0\rangle - |1\rangle) \otimes 1/\sqrt{2}(|0\rangle - |1\rangle) = |-\rangle| - \rangle.$$

In short, under the target qubit is $|-\rangle$, the control qubit becomes $|-\rangle$ from $|+\rangle$ by virtue of phase kickback.

2.10 Implementation of the phase oracle based on CNOT for the Bernstein–Vazirani algorithm—secret string $s = 101$ case

Suppose that

$$f: \{0, 1\}^3 \rightarrow \{0, 1\} \tag{2.10.1}$$

is a function with a 3-bit domain and a 1-bit range (figure 2.9). We assume the following function:

$$f(x) = s \cdot x = s_1x_1 \oplus s_2x_2 \oplus s_3x_3, \tag{2.10.2}$$

where $s_i \in \{0, 1\}$, $x_i \in \{0, 1\}$, $x = x_1x_2x_3 \in \{0, 1\}^3$, and $s = s_1s_2s_3 \in \{0, 1\}^3$. The goal is of determining $f(x)$. Here, the number of functions is 2^3. The Bernstein–Vazirani algorithm determines what function is true by one step. Thus, the efficiency of the Bernstein–Vazirani algorithm is 2^3.

For efficiency of operations in quantum oracle computing to the classical, we must note how many functions are used in an algorithm. Here, we sometimes see that the efficiency of this quantum operation seems to be N times the operation speed in classical operations, where N is the number of unknown bits which provide the secret string. However, thinking of the Bernstein–Vazirani algorithm-like quantum

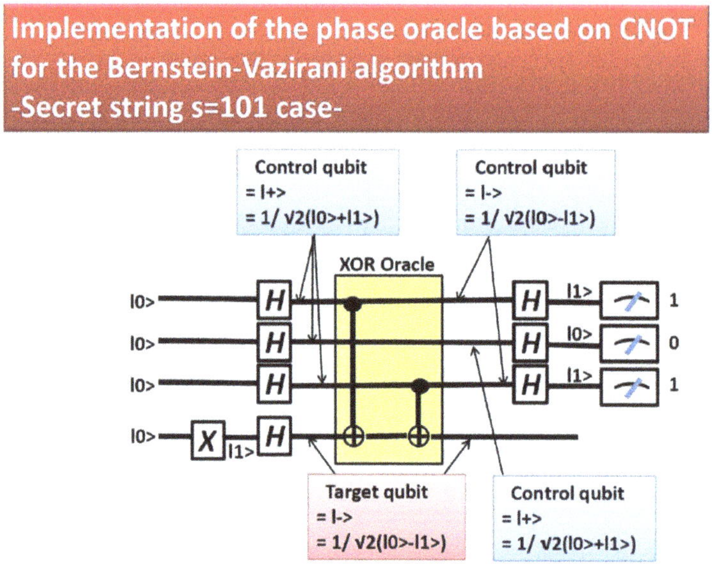

Figure 2.9. Implementation of the phase oracle based on CNOT for the Bernstein–Vazirani algorithm—secret string $s = 101$ case.

computing, we could see the efficiency appears to be 2^N times the operation speed of the classical. Briefly speaking, the efficiency depends upon the comparative object whenever we compare the two, which fact looks to be in our evaluations on comparison general.

We suppose $s_1 = 1$, $s_2 = 0$, $s_3 = 1$. Here, $x \oplus y$ is the bitwise XOR (exclusive OR) of x and y. Also, $s \cdot x$ is the bitwise inner product of s and x, modulo 2. The goal is of storing the string $s_1s_2s_3$ into an output quantum state as the coefficients of $f(x)$. Let us follow the quantum states through the Bernstein–Vazirani algorithm.

The input state is

$$|\psi_0\rangle = |0\rangle^{\otimes 3}|1\rangle. \tag{2.10.3}$$

After the componentwise Hadamard transformations on the input state, we have

$$|\psi_1\rangle = \sum_{x \in \{0,1\}^3} \frac{|x\rangle}{\sqrt{2^3}} \left[\frac{|0\rangle - |1\rangle}{\sqrt{2}} \right]. \tag{2.10.4}$$

Next, the function f is stored into a quantum state using

$$U_f|x, y\rangle = |x, y \oplus f(x)\rangle \tag{2.10.5}$$

in giving, by the phase kickback formation,

$$|\psi_2\rangle = \pm \sum_{x \in \{0,1\}^3} \frac{(-1)^{f(x)}|x\rangle}{\sqrt{2^3}} \left[\frac{|0\rangle - |1\rangle}{\sqrt{2}} \right]. \tag{2.10.6}$$

Here, we have

$$\sum_{x\in\{0,1\}^3} \frac{(-1)^{f(x)}|x\rangle}{\sqrt{2^3}} = \sum_{x\in\{0,1\}^3} \frac{(-1)^{s\cdot x}|x\rangle}{\sqrt{2^3}}$$
$$= \frac{1}{\sqrt{2}}(|0\rangle + (-1)^{s_1}|1\rangle) \otimes \frac{1}{\sqrt{2}}(|0\rangle + (-1)^{s_2}|1\rangle) \tag{2.10.7}$$
$$\otimes \frac{1}{\sqrt{2}}(|0\rangle + (-1)^{s_3}|1\rangle).$$

After the componentwise Hadamard transformations on the first three qubits in the state $|\psi_2\rangle$, we can now evaluate $|\psi_3\rangle$

$$|\psi_3\rangle = \pm|s_1s_2s_3\rangle\left[\frac{|0\rangle - |1\rangle}{\sqrt{2}}\right] = \pm|101\rangle\left[\frac{|0\rangle - |1\rangle}{\sqrt{2}}\right]. \tag{2.10.8}$$

Thus, in summary, we store $s_1s_2s_3$ in a single quantum state as $|101\rangle$ and the results of measurements are $s_1 = 1$, $s_2 = 0$, $s = 1$. According to the first supposition, we are succeeding in storing them correctly.

In the classical case we require 2^3 steps. In the quantum case here we require only one step.

Reference

[1] Nannicini G 2019 *Quantum Computing, Lecture 3* (Yorktown Heights, NY: IBM T.J. Watson)

IOP Publishing

Simplified Quantum Computing with Applications

Koji Nagata, Do Ngoc Diep, Ahmed Farouk and Tadao Nakamura

Chapter 3

Quantum key distribution based on a special Deutsch–Jozsa algorithm

This chapter provides quantum key distribution based on a special Deutsch–Jozsa algorithm.

3.1 Review of Deutsch's algorithm

We review Deutsch's algorithm as presented in Nielsen and Chuang.

Quantum superposition and phase kickback are fundamental features of many quantum algorithms. They allow quantum computers to evaluate simultaneously the values of a function $f(x)$ for many different x. Suppose

$$f\colon \{0, 1\} \rightarrow \{0, 1\} \tag{3.1.1}$$

is a function with a one-bit domain and range. A convenient way of computing the function on a quantum computer is to consider a two-qubit quantum computer that starts with the state $|x, y\rangle$, where x and y are variables used in mapping f. The abbreviation $|x, y\rangle$ (or $|x\rangle|y\rangle$ if any) stands for $|x\rangle \otimes |y\rangle$. With an appropriate sequence of logic gates, it is possible to transform this state into

$$|x, y \oplus f(x)\rangle, \tag{3.1.2}$$

where $\oplus$ indicates addition modulo 2. We denote by U_f the transformation defined by the mapping

$$U_f|x, y\rangle = |x, y \oplus f(x)\rangle. \tag{3.1.3}$$

Deutsch's algorithm combines quantum superposition with a property of quantum mechanics known as interference. Let us use the Hadamard gate to transform the first qubit $|0\rangle$ to the superposition

$$(|0\rangle + |1\rangle)/\sqrt{2} \tag{3.1.4}$$

doi:10.1088/978-0-7503-4700-6ch3

but let us transform the second qubit to the superposition

$$(|0\rangle - |1\rangle)/\sqrt{2} \tag{3.1.5}$$

using the Hadamard gate applied to the state $|1\rangle$. The Hadamard gate is given by

$$H = \frac{1}{\sqrt{2}}(|0\rangle\langle 1| + |1\rangle\langle 0| + |0\rangle\langle 0| - |1\rangle\langle 1|). \tag{3.1.6}$$

The matrix of it is given by

$$H = \frac{1}{\sqrt{2}}\begin{pmatrix} 1 & 1 \\ 1 & -1 \end{pmatrix}. \tag{3.1.7}$$

Let us follow the evolution of the states to see what happens in this circuit. The input state

$$|\psi_0\rangle = |01\rangle \tag{3.1.8}$$

is sent through two Hadamard gates to give

$$|\psi_1\rangle = \left[\frac{|0\rangle + |1\rangle}{\sqrt{2}}\right]\left[\frac{|0\rangle - |1\rangle}{\sqrt{2}}\right]. \tag{3.1.9}$$

The phase kickback formation shows that we apply U_f to the state

$$|x\rangle(|0\rangle - |1\rangle)/\sqrt{2} \tag{3.1.10}$$

then we obtain the state

$$(-1)^{f(x)}|x\rangle(|0\rangle - |1\rangle)/\sqrt{2}, \tag{3.1.11}$$

where we regard $\frac{|0\rangle+|1\rangle}{\sqrt{2}}$ as $|x\rangle$.

Generally speaking, so long as it is using $\left[\frac{|0\rangle-|1\rangle}{\sqrt{2}}\right]$ as its eigenstate for a gate, $|x\rangle$ of first qubit is changed into the $(-1)^{f(x)}|x\rangle$ that has the phase kickback factor $(-1)^{f(x)}$ after operation by the gate.

Applying U_f to $|\psi_1\rangle$ therefore leaves us with one of the two possibilities

$$|\psi_2\rangle = \begin{cases} \pm\left[\frac{|0\rangle + |1\rangle}{\sqrt{2}}\right]\left[\frac{|0\rangle - |1\rangle}{\sqrt{2}}\right] & \text{if } f(0) = f(1), \\ \pm\left[\frac{|0\rangle - |1\rangle}{\sqrt{2}}\right]\left[\frac{|0\rangle - |1\rangle}{\sqrt{2}}\right] & \text{if } f(0) \neq f(1). \end{cases} \tag{3.1.12}$$

According to the definition of the transformation U_f, the second qubit has the effect of it. The phase kickback formation transforms the effect from the second qubit into the first qubit. Finally, only the first qubit has the effect of the transformation U_f.

The final Hadamard gate on the qubits thus gives us

$$|\psi_3\rangle = \begin{cases} \pm|0\rangle|1\rangle & \text{if } f(0) = f(1), \\ \pm|1\rangle|1\rangle & \text{if } f(0) \neq f(1). \end{cases} \tag{3.1.13}$$

So, by measuring the first qubit, we may determine $f(x)$ is constant or balanced, which is obviously evaluated through the term $f(0) \oplus f(1)$. This is very interesting indeed: the quantum circuit gives us the ability to determine a global property of $f(x)$, namely, $f(0) \oplus f(1)$, using only one evaluation of $f(x)$! This is faster than is possible with a classical apparatus which would require at least two evaluations.

Later, the same situation can be applied to N-variable Boolean function case f: $\{0, 1\}^N \to \{0, 1\}$. In the case, classically one needs $2^{N-1} + 1$ operations to determine whether the function is constant or balanced. In our quantum situation we need only one measurement by virtue of one time evaluation/operation. The fact is, for example, applied to the problem of quantum secret sharing.

3.2 Deutsch's algorithm with another input state

We now study Deutsch's algorithm with another input state, for which it turns out that the algorithm cannot be implemented. The input state

$$|\psi_0\rangle = |10\rangle \tag{3.2.1}$$

is sent through two Hadamard gates to give

$$|\psi_1\rangle = \left[\frac{|0\rangle - |1\rangle}{\sqrt{2}}\right]\left[\frac{|0\rangle + |1\rangle}{\sqrt{2}}\right]. \tag{3.2.2}$$

We apply U_f to the state

$$\frac{|0\rangle - |1\rangle}{\sqrt{2}}|x\rangle. \tag{3.2.3}$$

If $x = 1$, the state is

$$\frac{|0\rangle|1\rangle - |1\rangle|1\rangle}{\sqrt{2}} \tag{3.2.4}$$

and the application of U_f gives

$$\frac{|0\rangle|\overline{f(0)}\rangle - |1\rangle|\overline{f(1)}\rangle}{\sqrt{2}}, \tag{3.2.5}$$

while if $x = 0$, the state is

$$\frac{|0\rangle|0\rangle - |1\rangle|0\rangle}{\sqrt{2}} \tag{3.2.6}$$

and the application of U_f gives

$$\frac{|0\rangle|f(0)\rangle - |1\rangle|f(1)\rangle}{\sqrt{2}}. \tag{3.2.7}$$

Thus, we have the superposition state

$$\frac{|0\rangle(|f(0)\rangle + |\overline{f(0)}\rangle) - |1\rangle(|f(1)\rangle + |\overline{f(1)}\rangle)}{\sqrt{2}}. \tag{3.2.8}$$

Applying U_f to $|\psi_1\rangle$ therefore leaves us with one of the two possibilities

$$|\psi_2\rangle = \begin{cases} \pm\left[\frac{|0\rangle - |1\rangle}{\sqrt{2}}\right]\left[\frac{|0\rangle + |1\rangle}{\sqrt{2}}\right] & \text{if } f(0) = f(1), \\ \pm\left[\frac{|0\rangle - |1\rangle}{\sqrt{2}}\right]\left[\frac{|0\rangle + |1\rangle}{\sqrt{2}}\right] & \text{if } f(0) \neq f(1). \end{cases} \tag{3.2.9}$$

The final Hadamard gate on the qubits thus gives us

$$|\psi_3\rangle = \begin{cases} \pm|1\rangle|0\rangle & \text{if } f(0) = f(1), \\ \pm|1\rangle|0\rangle & \text{if } f(0) \neq f(1). \end{cases} \tag{3.2.10}$$

In the case, the application of Deutsch's algorithm cannot be possible.

3.3 Deutsch's algorithm using the Bell state

We now study Deutsch's algorithm using the Bell state. The input state

$$|\psi_0\rangle = \frac{|10\rangle + |01\rangle}{\sqrt{2}} \tag{3.3.1}$$

is sent through two Hadamard gates to give

$$|\psi_1\rangle = \frac{1}{\sqrt{2}}\left(\left[\frac{|0\rangle - |1\rangle}{\sqrt{2}}\right]\left[\frac{|0\rangle + |1\rangle}{\sqrt{2}}\right] + \left[\frac{|0\rangle + |1\rangle}{\sqrt{2}}\right]\left[\frac{|0\rangle - |1\rangle}{\sqrt{2}}\right]\right). \tag{3.3.2}$$

Applying U_f to $|\psi_1\rangle$ therefore leaves us with one of the two possibilities

$$\begin{aligned} |\psi_2\rangle &= \pm\frac{1}{\sqrt{2}}\left(\left[\frac{|0\rangle - |1\rangle}{\sqrt{2}}\right]\left[\frac{|0\rangle + |1\rangle}{\sqrt{2}}\right] \pm \left[\frac{|0\rangle + |1\rangle}{\sqrt{2}}\right]\left[\frac{|0\rangle - |1\rangle}{\sqrt{2}}\right]\right) \\ &\quad \text{if } f(0) = f(1), \\ |\psi_2\rangle &= \pm\frac{1}{\sqrt{2}}\left(\left[\frac{|0\rangle - |1\rangle}{\sqrt{2}}\right]\left[\frac{|0\rangle + |1\rangle}{\sqrt{2}}\right] \pm \left[\frac{|0\rangle - |1\rangle}{\sqrt{2}}\right]\left[\frac{|0\rangle - |1\rangle}{\sqrt{2}}\right]\right) \\ &\quad \text{if } f(0) \neq f(1). \end{aligned} \tag{3.3.3}$$

The final Hadamard gate on the qubits thus gives us $|\psi_3\rangle$ as follows:

$$|\psi_3\rangle = \begin{cases} \pm\frac{|1\rangle|0\rangle \pm |0\rangle|1\rangle}{\sqrt{2}} & \text{if } f(0) = f(1) \quad \text{(entanglement)}, \\ \pm\frac{|1\rangle|0\rangle \pm |1\rangle|1\rangle}{\sqrt{2}} & \text{if } f(0) \neq f(1) \quad \text{(separable)}. \end{cases} \tag{3.3.4}$$

So, by measuring the qubits (by means of the Bell measurement), we can determine $f(0) \oplus f(1)$. The Bell measurement can be explained as follows. Alice and Bob prepare the Bell basis

$$|\Psi_+\rangle = \frac{|1\rangle|0\rangle + |0\rangle|1\rangle}{\sqrt{2}},$$
$$|\Psi_-\rangle = \frac{|1\rangle|0\rangle - |0\rangle|1\rangle}{\sqrt{2}},$$
$$|\Phi_+\rangle = \frac{|1\rangle|1\rangle + |0\rangle|0\rangle}{\sqrt{2}},$$
$$|\Phi_-\rangle = \frac{|1\rangle|1\rangle - |0\rangle|0\rangle}{\sqrt{2}}. \tag{3.3.5}$$

If the state $|\psi_3\rangle$ is an entangled state, we have

$$|\langle\psi_3|\Psi_+\rangle|^2 = 1,\ |\langle\psi_3|\Psi_-\rangle|^2 = 1,\ |\langle\psi_3|\Phi_+\rangle|^2 = 1, \text{ or } |\langle\psi_3|\Phi_-\rangle|^2 = 1. \tag{3.3.6}$$

Therefore, the measurement outcome should be 1 if the function is constant. If the state $|\psi_3\rangle$ is a separable state, we have

$$|\langle\psi_3|\Psi_+\rangle|^2 = \frac{1}{2},\ |\langle\psi_3|\Psi_-\rangle|^2 = \frac{1}{2},\ |\langle\psi_3|\Phi_+\rangle|^2 = \frac{1}{2}, \text{ or } |\langle\psi_3|\Phi_-\rangle|^2 = \frac{1}{2}. \tag{3.3.7}$$

Therefore, the measurement outcome should not be 1 if the function is balanced.

3.4 Quantum key distribution based on Deutsch's algorithm

We now discuss how Deutsch's algorithm can be used for a secure quantum key distribution using an entangled state. Let us assume that Alice and Bob promise to use a function f that is of one of two kinds: either the value of f is constant or it is balanced. To Eve, it is secret. Alice's and Bob's goal is to determine with certainty whether they chose a constant or a balanced function without providing information about the function to Eve. If the function is constant, the output qubits are entangled; otherwise they are separable. Alice and Bob perform the Bell measurement. They share one secret bit if they determine the function f by obtaining a suitable measurement outcome. The existence of Eve destroys entanglement. The security of our method is based on the Ekert 91 protocol.

- First, Alice prepares the entangled qubits, applies the Hadamard transformation to the state, and sends the output state described in the Bell state to Bob.
- Next, Bob randomly picks a function f that is either balanced or constant, and he applies U_f. He then sends the one qubit to Alice.
- Finally, Alice and Bob perform the Bell measurement. She learns whether f is balanced or constant. If the final qubits are entangled, then the function is constant. If the final qubits are not entangled, then the function is balanced. Alice and Bob now share a secret bit of information (the 'type' of $f(x)$).

- The result of the Bell measurement is 1 if the function is constant.
- Alice and Bob compare a subset of all the results of the Bell measurements when the function is constant; all of them should be 1.
- The existence of Eve must destroy entanglement (Ekert 91).
- Eve is detected in the case that the result of the Bell measurement is not 1 and the function is constant.

We thus show that Deutsch's algorithm can be used for the secure quantum key distribution with security based on the Ekert 91 protocol.

3.5 Review of the Deutsch–Jozsa algorithm

The earliest quantum algorithm, the Deutsch–Jozsa algorithm, is a representative example showing that quantum computation is faster than its classical counterpart by a factor that grows exponentially with the number of qubits. Let us follow the line of argument presented in Nielsen and Chuang. The application of the algorithm, known as *Deutsch's problem*, may be described in terms of the following game. Alice, in Amsterdam, selects a number x from 0 to $2^N - 1$, and mails it in a letter to Bob, in Boston. Bob calculates the value of some function

$$f: \{0, \dots, 2^N - 1\} \rightarrow \{0, 1\} \tag{3.5.1}$$

and replies with the result, which is either 0 or 1. Now, Bob promises to use a function f that is one of two kinds; either the value of $f(x)$ is constant for all x or it is balanced, i.e., it is equal to 1 for exactly half of the possible x and to 0 for the other half. Alice's goal is to determine with certainty whether Bob chooses a constant or a balanced function, corresponding with him as little as possible. How fast can she succeed in doing this?

In the classical case, Alice may only send Bob one value of x in each letter. At worst, Alice will need to query Bob at least

$$2^N/2 + 1 \tag{3.5.2}$$

times, since she may receive $2^N/2$ 0's before finally getting a 1, telling her that Bob's function is balanced. The best deterministic classical algorithm that she can use therefore requires $2^N/2 + 1$ queries. Note that in each letter, Alice sends Bob N bits of information. Furthermore, in this example, physical distance is being used to artificially elevate the cost of calculating $f(x)$, but this is not needed in the general problem, where $f(x)$ may be inherently difficult to calculate.

If Bob and Alice are able to exchange qubits, instead of just classical bits, and if Bob agrees to calculate $f(x)$ using a unitary transformation U_f, then Alice could achieve her goal in just one correspondence with Bob, using the following algorithm.

Alice has an N-qubit register to store her query in and a single-qubit register that she will give to Bob to store the answer in. She begins by preparing both her query and answer registers in a superposition state. Bob will evaluate $f(x)$ using quantum superposition and leave the result in the answer register. Alice then interferes with states in the superposition using a Hadamard transformation (a unitary transformation)

$$H = (\sigma_x + \sigma_z)/\sqrt{2} \tag{3.5.3}$$

on the query register, and finishes by performing a suitable measurement to determine whether f was constant or balanced.

Let us follow the quantum states through this algorithm. The input state is

$$|\psi_0\rangle = |0\rangle^{\otimes N}|1\rangle. \tag{3.5.4}$$

Here the query register describes the state of N qubits all prepared in this $|0\rangle$ state. After the componentwise Hadamard transformations on the query register and the Hadamard gate on the answer register, we have

$$|\psi_1\rangle = \left(\sum_{x=0}^{2^N-1} \frac{|x\rangle}{\sqrt{2^N}}\right)\left[\frac{|0\rangle - |1\rangle}{\sqrt{2}}\right]. \tag{3.5.5}$$

The query register is now a superposition of all values, and the answer register is in an evenly weighted superposition of $|0\rangle$ and $|1\rangle$. Next, the function f is evaluated using

$$U_f|x, y\rangle = |x, y \oplus f(x)\rangle \tag{3.5.6}$$

in giving

$$|\psi_2\rangle = \left(\sum_{x=0}^{2^N-1} \frac{(-1)^{f(x)}|x\rangle}{\sqrt{2^N}}\right)\left[\frac{|0\rangle - |1\rangle}{\sqrt{2}}\right]. \tag{3.5.7}$$

Generally speaking, so long as using $\left[\frac{|0\rangle - |1\rangle}{\sqrt{2}}\right]$ as its eigenstate for a gate, $|x\rangle$ of first qubit is changed into the $(-1)^{f(x)}|x\rangle$ that has the phase kickback factor $(-1)^{f(x)}$ after operation by the gate.

In what follows, we discuss an example where $N = 3$ as follows:

$$\frac{|0\rangle + |1\rangle}{\sqrt{2}}\frac{|0\rangle + |1\rangle}{\sqrt{2}}\frac{|0\rangle + |1\rangle}{\sqrt{2}}\frac{|0\rangle - |1\rangle}{\sqrt{2}}. \tag{3.5.8}$$

Thus, we have

$$\frac{|000\rangle + |001\rangle + |010\rangle + |011\rangle + |100\rangle + |101\rangle + |110\rangle + |111\rangle}{\sqrt{2^3}}\frac{|0\rangle - |1\rangle}{\sqrt{2}}. \tag{3.5.9}$$

Thus, we have by the phase kickback

$$\begin{aligned}&\frac{1}{\sqrt{2^3}}((-1)^{f(000)}|000\rangle + (-1)^{f(001)}|001\rangle + (-1)^{f(010)}|010\rangle + (-1)^{f(011)}|011\rangle\\ &+(-1)^{f(100)}|100\rangle + (-1)^{f(101)}|101\rangle + (-1)^{f(110)}|110\rangle + (-1)^{f(111)}|111\rangle)\\ &\frac{|0\rangle - |1\rangle}{\sqrt{2}}.\end{aligned} \tag{3.5.10}$$

Here $y \oplus f(x)$ is the bitwise XOR (exclusive OR) of y and $f(x)$. Alice now has a set of qubits in which the result of Bob's function evaluation is stored in the amplitude of the qubit superposition state. She now interferes in terms in the superposition using the componentwise Hadamard transformation (later we use it for short, the Hadamard transformation) on the query register. To determine the result of the Hadamard transformation, it helps to first calculate the effect of the Hadamard transformation on a state $|x\rangle$. By checking the cases $x = 0$ and $x = 1$ separately, we see that for a single qubit

$$H|x\rangle = \sum_z (-1)^{xz}|z\rangle/\sqrt{2}. \tag{3.5.11}$$

Thus, we have

$$H^{\otimes N}|x_1, \ldots, x_N\rangle = \frac{\sum_{z_1,\ldots,z_N} (-1)^{x_1z_1+\cdots+x_Nz_N}|z_1, \ldots, z_N\rangle}{\sqrt{2^N}}. \tag{3.5.12}$$

This can be summarized more succinctly in the very useful equation

$$H^{\otimes N}|x\rangle = \frac{\sum_z (-1)^{x\cdot z}|z\rangle}{\sqrt{2^N}}, \tag{3.5.13}$$

where $x \cdot z$ is the bitwise inner product of x and z, modulo 2. Using (3.5.13) and (3.5.7), we can now evaluate $|\psi_3\rangle$

$$|\psi_3\rangle = \left(\sum_{z=0}^{2^N-1}\sum_{x=0}^{2^N-1} \frac{(-1)^{x\cdot z+f(x)}|z\rangle}{2^N}\right)\left[\frac{|0\rangle - |1\rangle}{\sqrt{2}}\right]. \tag{3.5.14}$$

Alice now observes the query register. Note that the absolute value of the amplitude for the state $|0\rangle^{\otimes N}$ is

$$\sum_{x=0}^{2^N-1} (-1)^{f(x)}/2^N. \tag{3.5.15}$$

Let us look at the two possible cases, namely, f constant or f balanced, to discern what happens. In the case where f is constant, the absolute value of the amplitude for $|0\rangle^{\otimes N}$ is +1. Because $|\psi_3\rangle$ is of unit length, it follows that all the other amplitudes must be zero, and an observation will yield 0 for all N qubits in the query register. Thus, all the results of measurements are 0. If f is balanced, then the positive and negative contributions to the absolute value of the amplitude for $|0\rangle^{\otimes N}$ cancel, leaving an amplitude of zero, and a measurement must yield a result other than 0 (i.e., +1) on at least one qubit in the query register. In summary, if Alice measures all 0's, that is, all the results of measurements are 0, then the function is constant; otherwise the function is balanced.

3.6 Special Deutsch–Jozsa algorithm

Originally, Bob promises to use a function f that is one of two kinds: either the value of $f(x)$ is constant for all x or the value of $f(x)$ is balanced, i.e., it is equal to 1 for exactly half of all x and 0 for the other half. Here, Bob uses a special function when it is not constant. We may say the value of $f(x)$ is special.

We have the following when the value of $f(x)$ is special

$$\sum_{x\in\{0,1\}^N} \frac{(-1)^{f(x)}|x\rangle}{\sqrt{2^N}} = \left[\frac{|0\rangle - |1\rangle}{\sqrt{2}}\right]^{\otimes N}. \tag{3.6.1}$$

That is, the function has the following form

$$f(x) = \begin{cases} 0 & \text{if the number of 1's is even in } x, \\ +1 & \text{if the number of 1's is odd in } x. \end{cases} \tag{3.6.2}$$

Alice's goal is to determine with certainty whether Bob chooses a constant or a special function, corresponding with him as little as possible.

The input state

$$|\psi_0\rangle = |0\rangle^{\otimes N}|1\rangle \tag{3.6.3}$$

is sent through the componentwise Hadamard gates to give

$$|\psi_1\rangle = \left[\frac{|0\rangle + |1\rangle}{\sqrt{2}}\right]^{\otimes N}\left[\frac{|0\rangle - |1\rangle}{\sqrt{2}}\right]. \tag{3.6.4}$$

We apply U_f, thereby obtaining the following state

$$|\psi_2\rangle = \pm\sum_x \frac{(-1)^{f(x)}|x\rangle}{\sqrt{2^N}}\left[\frac{|0\rangle - |1\rangle}{\sqrt{2}}\right]. \tag{3.6.5}$$

Applying U_f (a unitary operation) to $|\psi_1\rangle$ therefore leaves us with one of the two possibilities

$$|\psi_2\rangle = \begin{cases} \pm\left[\frac{|0\rangle + |1\rangle}{\sqrt{2}}\right]^{\otimes N}\left[\frac{|0\rangle - |1\rangle}{\sqrt{2}}\right] & \text{if } f(x) = \text{constant}, \\ \pm\left[\frac{|0\rangle - |1\rangle}{\sqrt{2}}\right]^{\otimes N}\left[\frac{|0\rangle - |1\rangle}{\sqrt{2}}\right] & \text{if } f(x) = \text{special}. \end{cases} \tag{3.6.6}$$

The final Hadamard gate on the qubits thus gives us

$$|\psi_3\rangle = \begin{cases} \pm|0\rangle^{\otimes N}|1\rangle & \text{if } f(x) = \text{constant}, \\ \pm|1\rangle^{\otimes N}|1\rangle & \text{if } f(x) = \text{special}. \end{cases} \tag{3.6.7}$$

Hence, in the case, we are able to perform the special Deutsch–Jozsa algorithm.

3.7 Special Deutsch–Jozsa algorithm with another input state

We now study the special Deutsch–Jozsa algorithm with another input state, for which it turns out that the algorithm cannot be implemented. The input state

$$|\psi_0\rangle = |1\rangle^{\otimes N}|0\rangle \tag{3.7.1}$$

is sent through the componentwise Hadamard gates to give

$$|\psi_1\rangle = \left[\frac{|0\rangle - |1\rangle}{\sqrt{2}}\right]^{\otimes N}\left[\frac{|0\rangle + |1\rangle}{\sqrt{2}}\right]. \tag{3.7.2}$$

We apply U_f to the state

$$\frac{\overbrace{|a\rangle - |b\rangle + |c\rangle - \cdots + |d\rangle}^{2^N}}{\sqrt{2^N}}|x\rangle. \tag{3.7.3}$$

If $x = 1$, the state is

$$\frac{|a\rangle|1\rangle - |b\rangle|1\rangle + |c\rangle|1\rangle - \cdots + |d\rangle|1\rangle}{\sqrt{2^N}} \tag{3.7.4}$$

and the application of U_f gives

$$\frac{|a\rangle|\overline{f(a)}\,\rangle - |b\rangle|\overline{f(b)}\,\rangle + |c\rangle|\overline{f(c)}\,\rangle - \cdots + |d\rangle|\overline{f(d)}\,\rangle}{\sqrt{2^N}}, \tag{3.7.5}$$

while if $x = 0$, the state is

$$\frac{|a\rangle|0\rangle - |b\rangle|0\rangle + |c\rangle|0\rangle - \cdots + |d\rangle|0\rangle}{\sqrt{2^N}} \tag{3.7.6}$$

and the application of U_f gives

$$\frac{|a\rangle|f(a)\rangle - |b\rangle|f(b)\rangle + |c\rangle|f(c)\rangle - \cdots + |d\rangle|f(d)\rangle}{\sqrt{2^N}}. \tag{3.7.7}$$

Thus, we have the superposition state

$$\begin{aligned}&\frac{|a\rangle(|f(a)\rangle + |\overline{f(a)}\,\rangle) - |b\rangle(|f(b)\rangle + |\overline{f(b)}\,\rangle)}{\sqrt{2^N}}\\ &+\frac{|c\rangle(|f(c)\rangle + |\overline{f(c)}\,\rangle) - \cdots + |d\rangle(|f(d)\rangle + |\overline{f(d)}\,\rangle)}{\sqrt{2^N}}.\end{aligned} \tag{3.7.8}$$

Applying U_f to $|\psi_1\rangle$ therefore leaves us with one of the two possibilities

$$|\psi_2\rangle = \begin{cases} \pm\left[\frac{|0\rangle - |1\rangle}{\sqrt{2}}\right]^{\otimes N}\left[\frac{|0\rangle + |1\rangle}{\sqrt{2}}\right] & \text{if } f(x) = \text{constant}, \\ \pm\left[\frac{|0\rangle - |1\rangle}{\sqrt{2}}\right]^{\otimes N}\left[\frac{|0\rangle + |1\rangle}{\sqrt{2}}\right] & \text{if } f(x) = \text{special}. \end{cases} \tag{3.7.9}$$

The final Hadamard gate on the qubits thus gives us

$$|\psi_3\rangle = \begin{cases} \pm |1\rangle^{\otimes N}|0\rangle & \text{if } f(x) = \text{constant}, \\ \pm |1\rangle^{\otimes N}|0\rangle & \text{if } f(x) = \text{special}. \end{cases} \tag{3.7.10}$$

Hence, in the case, we are unable to perform the special Deutsch–Jozsa algorithm.

3.8 Special Deutsch–Jozsa algorithm using the GHZ state

The input state

$$|\psi_0\rangle = \frac{|1\rangle^{\otimes N}|0\rangle + |0\rangle^{\otimes N}|1\rangle}{\sqrt{2}} \tag{3.8.1}$$

is sent through the componentwise Hadamard gates to give

$$|\psi_1\rangle = \frac{1}{\sqrt{2}}\left(\left[\frac{|0\rangle - |1\rangle}{\sqrt{2}}\right]^{\otimes N}\left[\frac{|0\rangle + |1\rangle}{\sqrt{2}}\right] + \left[\frac{|0\rangle + |1\rangle}{\sqrt{2}}\right]^{\otimes N}\left[\frac{|0\rangle - |1\rangle}{\sqrt{2}}\right]\right). \tag{3.8.2}$$

Applying U_f to $|\psi_1\rangle$ therefore leaves us with one of the two possibilities

$$\begin{aligned} |\psi_2\rangle &= \pm\frac{1}{\sqrt{2}}\left(\left[\frac{|0\rangle - |1\rangle}{\sqrt{2}}\right]^{\otimes N}\left[\frac{|0\rangle + |1\rangle}{\sqrt{2}}\right] \pm \left[\frac{|0\rangle + |1\rangle}{\sqrt{2}}\right]^{\otimes N}\left[\frac{|0\rangle - |1\rangle}{\sqrt{2}}\right]\right) \\ &\quad \text{if } f(x) = \text{constant}, \\ |\psi_2\rangle &= \pm\frac{1}{\sqrt{2}}\left(\left[\frac{|0\rangle - |1\rangle}{\sqrt{2}}\right]^{\otimes N}\left[\frac{|0\rangle + |1\rangle}{\sqrt{2}}\right] \pm \left[\frac{|0\rangle - |1\rangle}{\sqrt{2}}\right]^{\otimes N}\left[\frac{|0\rangle - |1\rangle}{\sqrt{2}}\right]\right) \\ &\quad \text{if } f(x) = \text{special}. \end{aligned} \tag{3.8.3}$$

The final Hadamard gate on the qubits thus gives us

$$|\psi_3\rangle = \begin{cases} \pm\dfrac{|1\rangle^{\otimes N}|0\rangle \pm |0\rangle^{\otimes N}|1\rangle}{\sqrt{2}} & \text{if } f(x) = \text{constant} \quad \text{(GHZ)}, \\ \pm\dfrac{|1\rangle^{\otimes N}|0\rangle \pm |1\rangle^{\otimes N}|1\rangle}{\sqrt{2}} & \text{if } f(x) = \text{special} \quad \text{(separable)}. \end{cases} \tag{3.8.4}$$

So, by measuring the qubits (by means of the GHZ measurement), we can determine $f(x)$ is constant or special. The GHZ measurement is explained as follows. Alice and Bob prepare the following GHZ basis

$$\begin{aligned} |\Psi_+\rangle &= \frac{|1\rangle^{\otimes N}|0\rangle + |0\rangle^{\otimes N}|1\rangle}{\sqrt{2}}, \\ |\Psi_-\rangle &= \frac{|1\rangle^{\otimes N}|0\rangle - |0\rangle^{\otimes N}|1\rangle}{\sqrt{2}}, \\ |\Phi_+\rangle &= \frac{|1\rangle^{\otimes N}|1\rangle + |0\rangle^{\otimes N}|0\rangle}{\sqrt{2}}, \\ |\Phi_-\rangle &= \frac{|1\rangle^{\otimes N}|1\rangle - |0\rangle^{\otimes N}|0\rangle}{\sqrt{2}}. \end{aligned} \tag{3.8.5}$$

If the state $|\psi_3\rangle$ is the GHZ state, we have

$$|\langle\psi_3|\Psi_+\rangle|^2 = 1, \quad |\langle\psi_3|\Psi_-\rangle|^2 = 1, \quad |\langle\psi_3|\Phi_+\rangle|^2 = 1, \text{ or } |\langle\psi_3|\Phi_-\rangle|^2 = 1. \tag{3.8.6}$$

Therefore, the measurement outcome should be 1 if the function is constant. If the state $|\psi_3\rangle$ is a separable state, we have

$$|\langle\psi_3|\Psi_+\rangle|^2 = \frac{1}{2}, \quad |\langle\psi_3|\Psi_-\rangle|^2 = \frac{1}{2}, \quad |\langle\psi_3|\Phi_+\rangle|^2 = \frac{1}{2}, \text{ or } |\langle\psi_3|\Phi_-\rangle|^2 = \frac{1}{2}. \tag{3.8.7}$$

Therefore the measurement outcome should not be 1 if the function is special.

3.9 Quantum key distribution based on the special Deutsch–Jozsa algorithm

We now discuss how the special Deutsch–Jozsa algorithm can be used for a secure quantum key distribution using the GHZ state. Alice and Bob promise to use a function f that is of one of two kinds: either the value of f is constant or it is special. To Eve, it is secret. Alice's and Bob's goal is to determine with certainty whether they choose a constant or a special function without giving information about the function to Eve. If the function is constant, then the output qubits are fully entangled (the GHZ state); otherwise, there is a separable state. Alice and Bob perform the GHZ measurement mentioned above. They share one secret bit if they determine the function f by obtaining a suitable measurement outcome. Eve destroys the entangled state, giving a separable state. The security of our method is based on the Ekert 91 protocol.

- First, Alice prepares the entangled qubits, applies the Hadamard transformation to the state, and sends the output state described in the GHZ state to Bob.
- Next, Bob randomly picks a function f that is either special or constant, and he applies U_f. He then sends the values of f at N qubits to Alice.
- Finally, Alice and Bob perform the GHZ measurement. She learns whether f is special or constant. If the final qubits are entangled, then the function is constant. If the final qubits are not the GHZ state, then the function is special. Alice and Bob now share a secret bit of information (the 'type' of $f(x)$).
- The result of the GHZ measurement is 1 if the function is constant.
- Alice and Bob compare a subset of all the results of the GHZ measurements when the function is constant. All of them should be 1.
- Eve must destroy the GHZ state (Ekert 91).
- Eve is detected in the case that the result of the GHZ measurement is not 1 and the function is constant.

We thus show that the special Deutsch–Jozsa algorithm can be used for the secure quantum key distribution with security based on the Ekert 91 protocol. Our quantum key distribution is superior to its classical counterpart by a factor $O(2^N)$.

Parts of this chapter have been reprinted from Nagata K, Nakamura T and Farouk A 2017 Quantum cryptography based on the Deutsch–Jozsa algorithm *Int. J. Theor. Phys.* **56** 2887.

Koji Nagata, Do Ngoc Diep, Ahmed Farouk and Tadao Nakamura

Chapter 4

Quantum communication based on the Bernstein–Vazirani algorithm in a noisy environment

This chapter provides quantum communication based on the Bernstein–Vazirani algorithm in a noisy environment.

4.1 Review of the Bernstein–Vazirani algorithm

Suppose that

$$f\colon \{0, 1\}^N \to \{0, 1\} \tag{4.1.1}$$

is a function with an N-bit domain and a 1-bit range. We assume the following case:

$$f(x) = a \cdot x = \sum_{i=1}^{N} a_i x_i (\text{mod } 2) = a_1 x_1 \oplus a_2 x_2 \oplus a_3 x_3 \oplus \cdots \oplus a_N x_N, \tag{4.1.2}$$

where $a \in \{0, 1\}^N$. The goal is of determining $f(x)$. Here, the number of functions is 2^N. The Bernstein–Vazirani algorithm determines what function is true by one step. Thus, the efficiency of the Bernstein–Vazirani algorithm is 2^N. Let us follow the quantum states through the Bernstein–Vazirani algorithm. The input state is

$$|\psi_0\rangle = |0\rangle^{\otimes N} |1\rangle. \tag{4.1.3}$$

After the componentwise Hadamard transformations on the input state, we have

$$|\psi_1\rangle = \sum_{x \in \{0,1\}^N} \frac{|x\rangle}{\sqrt{2^N}} \left[\frac{|0\rangle - |1\rangle}{\sqrt{2}} \right]. \tag{4.1.4}$$

doi:10.1088/978-0-7503-4700-6ch4

Next, the function f is evaluated using

$$U_f|x, y\rangle = |x, y \oplus f(x)\rangle \tag{4.1.5}$$

in giving

$$|\psi_2\rangle = \pm\sum_x \frac{(-1)^{f(x)}|x\rangle}{\sqrt{2^N}}\left[\frac{|0\rangle - |1\rangle}{\sqrt{2}}\right]. \tag{4.1.6}$$

Here $y \oplus f(x)$ is the bitwise XOR (exclusive OR) of y and $f(x)$.

We have

$$H^{\otimes N}|x\rangle = \frac{\sum_z (-1)^{x\cdot z}|z\rangle}{\sqrt{2^N}}, \tag{4.1.7}$$

where $x \cdot z$ is the bitwise inner product of x and z, modulo 2.

Using (4.1.7) and (4.1.6), we can now evaluate $|\psi_3\rangle$

$$|\psi_3\rangle = \pm\sum_z\sum_x \frac{(-1)^{x\cdot z + f(x)}|z\rangle}{2^N}\left[\frac{|0\rangle - |1\rangle}{\sqrt{2}}\right]. \tag{4.1.8}$$

Thus, we have

$$|\psi_3\rangle = \pm\sum_z\sum_x \frac{(-1)^{x\cdot z + a\cdot x}|z\rangle}{2^N}\left[\frac{|0\rangle - |1\rangle}{\sqrt{2}}\right]. \tag{4.1.9}$$

Notice that

$$\sum_x (-1)^{x\cdot z + a\cdot x} = 2^N\delta_{a,z}. \tag{4.1.10}$$

Thus, we have

$$\begin{aligned}
|\psi_3\rangle &= \pm\sum_z\sum_x \frac{(-1)^{x\cdot z + a\cdot x}|z\rangle}{2^N}\left[\frac{|0\rangle - |1\rangle}{\sqrt{2}}\right] \\
&= \pm\sum_z \frac{2^N\delta_{a,z}|z\rangle}{2^N}\left[\frac{|0\rangle - |1\rangle}{\sqrt{2}}\right] \\
&= \pm\,|a\rangle\left[\frac{|0\rangle - |1\rangle}{\sqrt{2}}\right] \\
&= \pm\,|a_1a_2a_3\cdots a_N\rangle\left[\frac{|0\rangle - |1\rangle}{\sqrt{2}}\right].
\end{aligned} \tag{4.1.11}$$

Alice now observes $|a_1a_2a_3\cdots a_N\rangle$. In summary, if Alice measures $|a_1a_2a_3\cdots a_N\rangle$, then the function is

$$f(x_1, x_2, \ldots, x_N) = a_1x_1 \oplus a_2x_2 \oplus a_3x_3 \oplus \cdots \oplus a_Nx_N. \tag{4.1.12}$$

This shows that the quantum algorithm is superior to its classical counterpart by a factor of 2^N.

4.2 Quantum communication based on the Bernstein–Vazirani algorithm

Alice and Bob promise to select a function $f(x_1, x_2, \ldots, x_N) = a_1x_1 \oplus a_2x_2 \oplus a_3x_3 \oplus \cdots \oplus a_Nx_N$. Alice does not know $a_1, a_2, \ldots, a_N$, but Bob does. Alice's goal is of determining with certainty what $a_1, a_2, \ldots, a_N$ Bob chooses. In classical theory, Alice has to ask Bob 2^N questions. In quantum theory, Alice has to ask Bob just *one* question! Alice prepares a suitable $(N + 1)$-partite uncorrelated state, performs the componentwise Hadamard transformation on this state, and sends the output state to Bob. Bob then performs the Bernstein–Vazirani algorithm and inputs the information about the a's into the final state. Alice asks him what the state is. Alice measures the final state, and she knows the a's. If the a's are learned by Alice, then Alice and Bob share N bits of information by just one communication with each other. The speed to share N bits is improved by a factor of 2^N in comparison with the classical case. This shows that the quantum communication is superior to the classical case by a factor of 2^N.

- First, Alice prepares the qubits in state (4.1.4) and sends the $N + 1$ qubits to Bob.
- Next, Bob picks up N bits 'a' and applies U_f as transformation (4.1.5) to (4.1.4), thereby the state is evolving the $N + 1$ qubits to give (4.1.6). He then sends the N qubits to Alice.
- Finally, Alice applies the Hadamard transformation to each of the qubits, and performs a measurement. She learns $f(x) = a \cdot x = \sum_{i=1}^{N} a_i x_i (\text{mod } 2) = a_1x_1 \oplus a_2x_2 \oplus a_3x_3 \oplus \cdots \oplus a_Nx_N$. Alice and Bob now share N bits of information (the 'type' of $f(x)$).
- In the classical case (without the quantum computation), Alice needs 2^N communications with Bob to share N bits of information.

We thus show that quantum communication is superior to classical communication by a factor of 2^N when the Bernstein–Vazirani algorithm is used. However, there may be an error between Alice's bit-string and Bob's. In the next section, we discuss error correction using the Bernstein–Vazirani algorithm.

4.3 Error correction based on the Bernstein–Vazirani algorithm

Suppose that

$$f: \{0, 1\}^N \to \{0, 1\} \tag{4.3.1}$$

is a function with an N-bit domain and a 1-bit range. We introduce two functions $g(x)$ and $h(x)$ that are related to the function $f(x)$ by

$$f(x) = g(x) \oplus h(x). \tag{4.3.2}$$

We assume the following case

$$\begin{aligned} g(x) &= b \cdot x = \sum_{i=1}^{N} b_i x_i (\text{mod } 2) = b_1 x_1 \oplus b_2 x_2 \oplus b_3 x_3 \oplus \cdots \oplus b_N x_N, \\ h(x) &= c \cdot x = \sum_{i=1}^{N} c_i x_i (\text{mod } 2) = c_1 x_1 \oplus c_2 x_2 \oplus c_3 x_3 \oplus \cdots \oplus c_N x_N, \\ f(x) &= \sum_{i=1}^{N} (b_i \oplus c_i) x_i (\text{mod } 2) \\ &= (b_1 \oplus c_1) x_1 \oplus (b_2 \oplus c_2) x_2 \oplus (b_3 \oplus c_3) x_3 \oplus \cdots \oplus (b_N \oplus c_N) x_N, \end{aligned} \tag{4.3.3}$$

where $b_j, c_j = 0, 1,$ and $x_j = 0, 1$. After the communication process, Alice has the bit-string $b = (b_1, b_2, \ldots, b_N)$ and Bob has the bit-string $c = (c_1, c_2, \ldots, c_N)$. We want to correct errors between these.

Let us follow the quantum states through the algorithm. The input state is

$$|\psi_0\rangle = |0\rangle^{\otimes N} |1\rangle. \tag{4.3.4}$$

After the componentwise Hadamard transformations on the input state, we have

$$|\psi_1\rangle = \sum_{x \in \{0,1\}^N} \frac{|x\rangle}{\sqrt{2^N}} \left[\frac{|0\rangle - |1\rangle}{\sqrt{2}} \right]. \tag{4.3.5}$$

Next, the function f is evaluated using

$$U_f |x, y\rangle = |x, y \oplus f(x)\rangle \tag{4.3.6}$$

in giving

$$|\psi_2\rangle = \pm \sum_x \frac{(-1)^{f(x)} |x\rangle}{\sqrt{2^N}} \left[\frac{|0\rangle - |1\rangle}{\sqrt{2}} \right]. \tag{4.3.7}$$

After the Hadamard transformation, we can now evaluate $|\psi_3\rangle$ using (4.1.7) and (4.3.7)

$$|\psi_3\rangle = \pm \sum_z \sum_x \frac{(-1)^{x \cdot z + f(x)} |z\rangle}{2^N} \left[\frac{|0\rangle - |1\rangle}{\sqrt{2}} \right]. \tag{4.3.8}$$

We have

$$\begin{aligned} |\psi_3\rangle &= \pm \sum_z \sum_x \frac{(-1)^{x \cdot z + g(x) \oplus h(x)} |z\rangle}{2^N} \left[\frac{|0\rangle - |1\rangle}{\sqrt{2}} \right] \\ &= \pm \sum_z \sum_x \frac{(-1)^{x \cdot z + x \cdot b \oplus x \cdot c} |z\rangle}{2^N} \left[\frac{|0\rangle - |1\rangle}{\sqrt{2}} \right] \\ &= \pm \sum_z \sum_x \frac{(-1)^{x \cdot z + x \cdot (b + c)} |z\rangle}{2^N} \left[\frac{|0\rangle - |1\rangle}{\sqrt{2}} \right], \end{aligned} \tag{4.3.9}$$

where

$$b + c = (b_1 \oplus c_1, b_2 \oplus c_2, \ldots, b_N \oplus c_N). \tag{4.3.10}$$

Notice that

$$\sum_x (-1)^{x\cdot z + x\cdot(b+c)} = 2^N \delta_{(b+c),z}. \tag{4.3.11}$$

Thus, we have

$$\begin{aligned}
|\psi_3\rangle &= \pm \sum_z \sum_x \frac{(-1)^{x\cdot z + x\cdot(b+c)}|z\rangle}{2^N}\left[\frac{|0\rangle - |1\rangle}{\sqrt{2}}\right] \\
&= \pm \sum_z \frac{2^N \delta_{(b+c),z}|z\rangle}{2^N}\left[\frac{|0\rangle - |1\rangle}{\sqrt{2}}\right] \\
&= \pm |b + c\rangle\left[\frac{|0\rangle - |1\rangle}{\sqrt{2}}\right] \\
&= \pm |b_1 \oplus c_1, b_2 \oplus c_2, b_3 \oplus c_3, \ldots, b_N \oplus c_N\rangle\left[\frac{|0\rangle - |1\rangle}{\sqrt{2}}\right].
\end{aligned} \tag{4.3.12}$$

Alice now observes $|b_1 \oplus c_1, b_2 \oplus c_2, b_3 \oplus c_3, \ldots, b_N \oplus c_N\rangle$. In summary, if Alice measures $|100\cdots0\rangle$, the relation is

$$b_1 \oplus c_1 = 1, \quad b_2 \oplus c_2 = 0, \quad \ldots, \quad b_N \oplus c_N = 0. \tag{4.3.13}$$

Thus, there is an error for the first bit

$$b_1 \neq c_1, \quad b_2 = c_2, \quad \ldots, \quad b_N = c_N. \tag{4.3.14}$$

Hence, Alice detects the error. In general, Alice is able to determine the locations of such errors.

If Alice measures $|000\cdots0\rangle$, the relation is

$$b_1 \oplus c_1 = 0, \quad b_2 \oplus c_2 = 0, \quad \ldots, \quad b_N \oplus c_N = 0. \tag{4.3.15}$$

Thus, Alice and Bob share N bits of information

$$b_1 = c_1, \quad b_2 = c_2, \quad \ldots, \quad b_N = c_N. \tag{4.3.16}$$

We see that the quantum error correction is superior to its classical counterpart by a factor of 2^N in the case.

4.4 Evaluating simultaneously many functions using many parallel quantum systems

We now propose a method for evaluating many functions at the same time by using many parallel quantum systems together with the Bernstein–Vazirani algorithm. Given the set of real values $\{a_1, a_2, a_3, \ldots, a_N\}$ and a special function g: $\mathbf{R} \to \{0, 1\}$, we shall determine simultaneously the values (a bit-string) $g(a_1), g(a_2), g(a_3), \ldots, g(a_N)$. $\mathbf{R}$ is the set of real numbers. The speed of determining the values is shown to outperform the classical case by a factor of 2^N. Using M parallel quantum systems, we can evaluate simultaneously M functions $g^1, g^2, \ldots, g^M$. The speed of determining the $2^N \times M$ values will be shown to outperform the classical case by a factor of $2^N \times M$.

Let us suppose that we are given a sequence of real values $a_1, a_2, a_3, \ldots, a_N$, and let us introduce a function

$$g: \mathbf{R} \to \{0, 1\}. \tag{4.4.1}$$

Our first step is of determining the values

$$g(a_1), \quad g(a_2), \quad g(a_3), \quad \ldots, \quad g(a_N). \tag{4.4.2}$$

Recall that in the classical case, we need 2^N queries, i.e., 2^N separate evaluations of the function (4.4.1). In our quantum algorithm, we shall require a single query. Suppose now that we introduce another function

$$f: \{0, 1\}^N \to \{0, 1\} \tag{4.4.3}$$

which is a function with an N-bit domain and a 1-bit range. We construct the following function:

$$\begin{aligned} f(x) = g(a) \cdot x &= \sum_{i=1}^{N} g(a_i)x_i(\text{mod } 2) \\ &= g(a_1)x_1 \oplus g(a_2)x_2 \oplus g(a_3)x_3 \oplus \cdots \oplus g(a_N)x_N, \end{aligned} \tag{4.4.4}$$

where $x_i \in \{0, 1\}$, $g(a_i) \in \{0, 1\}$, and $a_i \in \mathbf{R}$. Here $g(a)$ symbolizes

$$g(a_1)g(a_2)\cdots g(a_N). \tag{4.4.5}$$

Let us follow the quantum states through the algorithm. The input state is

$$|\psi_0\rangle = |0\rangle^{\otimes N}|1\rangle. \tag{4.4.6}$$

After the componentwise Hadamard transformations on the input state, we have

$$|\psi_1\rangle = \sum_{x\in\{0,1\}^N} \frac{|x\rangle}{\sqrt{2^N}}\left[\frac{|0\rangle - |1\rangle}{\sqrt{2}}\right]. \tag{4.4.7}$$

Next, the function f is evaluated using

$$U_f|x, y\rangle = |x, y \oplus f(x)\rangle \tag{4.4.8}$$

in giving

$$|\psi_2\rangle = \pm\sum_{x} \frac{(-1)^{f(x)}|x\rangle}{\sqrt{2^N}}\left[\frac{|0\rangle - |1\rangle}{\sqrt{2}}\right]. \tag{4.4.9}$$

After the Hadamard transformation, using (4.1.7) and (4.4.9) we can now evaluate $|\psi_3\rangle$

$$|\psi_3\rangle = \pm\sum_{z}\sum_{x} \frac{(-1)^{x\cdot z+f(x)}|z\rangle}{2^N}\left[\frac{|0\rangle - |1\rangle}{\sqrt{2}}\right]. \tag{4.4.10}$$

Thus, we have

$$|\psi_3\rangle = \pm\sum_z\sum_x \frac{(-1)^{x\cdot z+g(a)\cdot x}|z\rangle}{2^N}\left[\frac{|0\rangle - |1\rangle}{\sqrt{2}}\right]. \tag{4.4.11}$$

Notice that

$$\sum_x(-1)^{x\cdot z+g(a)\cdot x} = 2^N\delta_{g(a),z}. \tag{4.4.12}$$

Thus, we have

$$\begin{aligned}|\psi_3\rangle &= \pm\sum_z\sum_x \frac{(-1)^{x\cdot z+g(a)\cdot x}|z\rangle}{2^N}\left[\frac{|0\rangle - |1\rangle}{\sqrt{2}}\right]\\ &= \pm\sum_z \frac{2^N\delta_{g(a),z}|z\rangle}{2^N}\left[\frac{|0\rangle - |1\rangle}{\sqrt{2}}\right]\\ &= \pm\,|g(a)\rangle\left[\frac{|0\rangle - |1\rangle}{\sqrt{2}}\right]\\ &= \pm\,|g(a_1)g(a_2)\cdots g(a_N)\rangle\left[\frac{|0\rangle - |1\rangle}{\sqrt{2}}\right]\end{aligned} \tag{4.4.13}$$

from which $|g(a_1)g(a_2)\cdots g(a_N)\rangle$ can be obtained. That is to say, if we measure $|g(a_1)g(a_2)\cdots g(a_N)\rangle$, then we can retrieve the values

$$g(a_1),\quad g(a_2),\quad g(a_3),\quad \dots,\quad g(a_N) \tag{4.4.14}$$

using a single query. All we have to do is of performing one quantum measurement. The speed to determine 2^N combinations of values is improved by a factor of 2^N compared with the classical counterpart. Notice that we recover the Bernstein–Vazirani algorithm when g: $a_i \rightarrow a_i$ and $a_i = 0, 1$.

We propose a method, using M parallel quantum systems, for evaluating simultaneously M functions $g^1, g^2, \dots, g^M$. That is, we can retrieve the values

$$\begin{aligned}&g^1(a_1),\, g^1(a_2),\, g^1(a_3),\, \dots,\, g^1(a_N),\\ &g^2(a_1),\, g^2(a_2),\, g^2(a_3),\, \dots,\, g^2(a_N),\\ &\vdots\\ &g^M(a_1),\, g^M(a_2),\, g^M(a_3),\, \dots,\, g^M(a_N).\end{aligned} \tag{4.4.15}$$

In this case, we measure the following quantum state

$$\begin{aligned}&|g^1(a_1)g^1(a_2)\cdots g^1(a_N)\rangle \otimes |g^2(a_1)g^2(a_2)\cdots g^2(a_N)\rangle\\ &\otimes\cdots\otimes|g^M(a_1)g^M(a_2)\cdots g^M(a_N)\rangle.\end{aligned} \tag{4.4.16}$$

All we have to do is perform one quantum measurement.

4.5 Method for evaluating a multiplication operation using the generalized Bernstein–Vazirani algorithm

We now present a new method for high-speed evaluation of a multiplication operation using the generalized Bernstein–Vazirani algorithm and many parallel quantum systems. Given the set of real values $\{a_1, a_2, a_3, \ldots, a_N\}$ and a special function g: $\mathbf{R} \rightarrow \{0, 1\}$, we shall evaluate simultaneously the values (a bit-string)

$$g(a_1), g(a_2), g(a_3), \ldots, g(a_N). \tag{4.5.1}$$

R is the set of real numbers. The speed of evaluating the values is shown to outperform the classical case by a factor of 2^N. We then consider a number in binary representation of it: $M_1 = (g(a_1), g(a_2), g(a_3), \ldots, g(a_N))$. By using M parallel quantum systems, we have simultaneously M numbers in binary representation. The speed of obtaining the M numbers is shown to outperform the classical case by a factor of M. Finally, we calculate the product $M_1 \times M_2 \times \cdots \times M_M$. The speed of evaluating the product is shown to outperform the classical case by a factor of $2^N \times M$.

We suppose that we are given a sequence of real values $a_1, a_2, a_3, \ldots, a_N$. Let us now introduce a function

$$g: \mathbf{R} \rightarrow \{0, 1\}. \tag{4.5.2}$$

Our first step is of determining the values

$$g(a_1), \quad g(a_2), \quad g(a_3), \quad \ldots, \quad g(a_N). \tag{4.5.3}$$

Recall that in the classical case, we need 2^N queries, i.e., 2^N separate evaluations of the function (4.5.2). In our quantum algorithm, we shall require just a single query. Suppose now that we introduce another function

$$f: \{0, 1\}^N \rightarrow \{0, 1\} \tag{4.5.4}$$

which is a function with an N-bit domain and a 1-bit range. We construct the following function:

$$f(x) = g(a) \cdot x = \sum_{i=1}^{N} g(a_i)x_i (\text{mod } 2) = g(a_1)x_1 \oplus g(a_2)x_2 \oplus g(a_3)x_3 \oplus \cdots \oplus g(a_N)x_N \tag{4.5.5}$$

where $x_i \in \{0, 1\}$, $g(a_i) \in \{0, 1\}$, and $a_i \in \mathbf{R}$. Here $g(a)$ symbolizes

$$g(a_1)g(a_2) \cdots g(a_N). \tag{4.5.6}$$

Let us follow the quantum states through the algorithm. The input state is

$$|\psi_0\rangle = |0\rangle^{\otimes N}|1\rangle. \tag{4.5.7}$$

After the componentwise Hadamard transformations on the input state, we have

$$|\psi_1\rangle = \sum_{x \in \{0,1\}^N} \frac{|x\rangle}{\sqrt{2^N}} \left[\frac{|0\rangle - |1\rangle}{\sqrt{2}} \right]. \tag{4.5.8}$$

Next, the function f is evaluated using

$$U_f|x, y\rangle = |x, y \oplus f(x)\rangle \tag{4.5.9}$$

in giving

$$|\psi_2\rangle = \pm \sum_x \frac{(-1)^{f(x)}|x\rangle}{\sqrt{2^N}} \left[\frac{|0\rangle - |1\rangle}{\sqrt{2}} \right]. \tag{4.5.10}$$

After the Hadamard transformation, using (4.1.7) and (4.5.10), we can now evaluate $|\psi_3\rangle$

$$|\psi_3\rangle = \pm \sum_z \sum_x \frac{(-1)^{x \cdot z + f(x)}|z\rangle}{2^N} \left[\frac{|0\rangle - |1\rangle}{\sqrt{2}} \right]. \tag{4.5.11}$$

Thus, we have

$$|\psi_3\rangle = \pm \sum_z \sum_x \frac{(-1)^{x \cdot z + g(a) \cdot x}|z\rangle}{2^N} \left[\frac{|0\rangle - |1\rangle}{\sqrt{2}} \right]. \tag{4.5.12}$$

Notice that

$$\sum_x (-1)^{x \cdot z + g(a) \cdot x} = 2^N \delta_{g(a),z}. \tag{4.5.13}$$

Thus, we have

$$\begin{aligned} |\psi_3\rangle &= \pm \sum_z \sum_x \frac{(-1)^{x \cdot z + g(a) \cdot x}|z\rangle}{2^N} \left[\frac{|0\rangle - |1\rangle}{\sqrt{2}} \right] \\ &= \pm \sum_z \frac{2^N \delta_{g(a),z}|z\rangle}{2^N} \left[\frac{|0\rangle - |1\rangle}{\sqrt{2}} \right] \\ &= \pm\, |g(a)\rangle \left[\frac{|0\rangle - |1\rangle}{\sqrt{2}} \right] \\ &= \pm\, |g(a_1)g(a_2) \cdots g(a_N)\rangle \left[\frac{|0\rangle - |1\rangle}{\sqrt{2}} \right] \end{aligned} \tag{4.5.14}$$

from which $|g(a_1)g(a_2) \cdots g(a_N)\rangle$ can be obtained. That is to say, if we measure $|g(a_1)g(a_2) \cdots g(a_N)\rangle$, then we can retrieve the values

$$g(a_1), \quad g(a_2), \quad g(a_3), \quad \ldots, \quad g(a_N) \tag{4.5.15}$$

using a single query. All we have to do is perform one quantum measurement. The speed of evaluating 2^N combinations of values is improved by a factor of 2^N

compared with the classical counterpart. Notice that we recover the Bernstein–Vazirani algorithm when g: $a_i \to a_i$ and $a_i = 0, 1$.

We present a new method for high-speed calculation of a multiplication operation using many parallel quantum systems. With M parallel quantum systems, we can evaluate simultaneously M functions $g^1, g^2, \ldots, g^M$. That is, we can retrieve the following values:

$$\begin{array}{l} g^1(a_1),\ g^1(a_2),\ g^1(a_3),\ \ldots, g^1(a_N), \\ g^2(b_1),\ g^2(b_2),\ g^2(b_3),\ \ldots, g^2(b_N), \\ \vdots \qquad \vdots \qquad \vdots \qquad \vdots \\ g^M(c_1), g^M(c_2), g^M(c_3), \ldots, g^M(c_N). \end{array} \tag{4.5.16}$$

In this case, we measure the following quantum state

$$\begin{array}{l} |g^1(a_1)g^1(a_2) \cdots g^1(a_N)\rangle \otimes |g^2(b_1)g^2(b_2) \cdots g^2(b_N)\rangle \\ \otimes \cdots \otimes |g^M(c_1)g^M(c_2) \cdots g^M(c_N)\rangle. \end{array} \tag{4.5.17}$$

All we have to do is perform one quantum measurement.

Using M parallel quantum systems, we have simultaneously the following M numbers in binary representation.

$$\begin{array}{l} M_1 = (g^1(a_1), g^1(a_2), g^1(a_3), \ldots, g^1(a_N)), \\ M_2 = (g^2(b_1), g^2(b_2), g^2(b_3), \ldots, g^2(b_N)), \\ \vdots \\ M_M = (g^M(c_1), g^M(c_2), g^M(c_3), \ldots, g^M(c_N)). \end{array} \tag{4.5.18}$$

The speed of obtaining these M numbers is shown to outperform the classical case by a factor of M.

Finally, we evaluate the product $M_1 \times M_2 \times \cdots \times M_M$. The speed of obtaining this product is shown to outperform the classical case by a factor of $2^N \times M$.

As an example, if $N = 2$ and $M = 3$, we have

$$\begin{array}{l} (g^1(a_1), g^1(a_2)) = (0, 1) = 1, \\ (g^2(b_1), g^2(b_2)) = (1, 0) = 2, \\ (g^3(c_1), g^3(c_2)) = (1, 1) = 3, \end{array} \tag{4.5.19}$$

and

$$(g^1(a_1), g^1(a_2)) \times (g^2(b_1), g^2(b_2)) \times (g^3(c_1), g^3(c_2)) = 2 \times 3 = 6. \tag{4.5.20}$$

4.6 Bernstein–Vazirani algorithm in a noisy environment

In this section, we discuss the Bernstein–Vazirani algorithm in a noisy environment. Suppose that

$$f: \{0, 1\}^N \to \{0, 1\} \tag{4.6.1}$$

is a noiseless function with an N-bit domain and a 1-bit range. We introduce a noisy function g using the function $f(x)$

$$g(x) = f(x) \pm O(\epsilon). \tag{4.6.2}$$

Here $O(\epsilon) \ll 1$ is the noise term.

The noise is explained as follows: suppose that two qubits are described by a superposition state and that the value of a function $f(1)$ has an error. Then there must be two error states. (For example, when we treat 100 bits and there are two errors, the error probability is 2/100=1/50.)

Let us explain this by using a quantum superposition state

$$|\psi\rangle = \frac{|1\rangle_1 + |0\rangle_1}{\sqrt{2}} \frac{|1\rangle_2 + |0\rangle_2}{\sqrt{2}} = \frac{|1\rangle_1|1\rangle_2 + |1\rangle_1|0\rangle_2 + |0\rangle_1|1\rangle_2 + |0\rangle_1|0\rangle_2}{2}. \tag{4.6.3}$$

The function f is evaluated using

$$U_f|x, y\rangle = |x, y \oplus f(x)\rangle. \tag{4.6.4}$$

Thus, we have

$$\begin{aligned} U_f|\psi\rangle &= U_f \frac{|1, 1\rangle + |1, 0\rangle + |0, 1\rangle + |0, 0\rangle}{2} \\ &= \frac{|1, 1 \oplus f(1)\rangle + |1, 0 \oplus f(1)\rangle + |0, 1 \oplus f(0)\rangle + |0, 0 \oplus f(0)\rangle}{2}. \end{aligned} \tag{4.6.5}$$

Therefore, there are two $f(1)$'s in the output state. If there is an error in $f(1)$, then the two states

$$|1, 1 \oplus f(1)\rangle, \quad |1, 0 \oplus f(1)\rangle \tag{4.6.6}$$

have simultaneously an error. Thus, the number of errors is even. Here, we treat globally such errors in a statistical model.

We assume the following case

$$\begin{aligned} g(x) &= a \cdot x = \sum_{i=1}^{N} a_i' x_i (\text{mod } 2) \pm O(\epsilon) \\ &= a_1' x_1 \oplus a_2' x_2 \oplus a_3' x_3 \oplus \cdots \oplus a_N' x_N \pm \epsilon(x_1 + x_2 \cdots + x_N) \\ &= f(x) \pm O(\epsilon), \end{aligned} \tag{4.6.7}$$

where $a_j = a_j' \pm \epsilon$, $a_j' = 0, 1$, and $x_j = 0, 1$. We want to evaluate simultaneously $a_1, a_2, \ldots, a_N$ with a success probability so that we evaluate the noisy function $g(x)$ with the success probability. This is the formulation of the Bernstein–Vazirani algorithm in a noisy environment.

Let us follow the quantum states through the algorithm. The input state is

$$|\psi_0\rangle = |0\rangle^{\otimes N}|1\rangle. \tag{4.6.8}$$

After the componentwise Hadamard transformations on the input state, we have

$$|\psi_1\rangle = \sum_{x\in\{0,1\}^N} \frac{|x\rangle}{\sqrt{2^N}}\left[\frac{|0\rangle - |1\rangle}{\sqrt{2}}\right]. \tag{4.6.9}$$

Next, the function g is approximately evaluated using

$$U_g|x, y\rangle = |x, y \oplus [g(x)]\rangle. \tag{4.6.10}$$

On the real line, $[g(x)]$ denotes the nearest natural number to $g(x)$. Here, we see that $[g(x)] = 0, 1$ and

$$[g(x)] = f(x). \tag{4.6.11}$$

We have

$$|\psi_2\rangle = \pm\sum_x \frac{(-1)^{[g(x)]}|x\rangle}{\sqrt{2^N}}\left[\frac{|0\rangle - |1\rangle}{\sqrt{2}}\right]. \tag{4.6.12}$$

After the Hadamard transformation, we can now evaluate $|\psi_3\rangle$ using (4.1.7) and (4.6.12)

$$|\psi_3\rangle = \pm\sum_z\sum_x \frac{(-1)^{x\cdot z+[g(x)]}|z\rangle}{2^N}\left[\frac{|0\rangle - |1\rangle}{\sqrt{2}}\right]. \tag{4.6.13}$$

So we have

$$|\psi_3\rangle = \pm\sum_z\sum_x \frac{(-1)^{x\cdot z+g(x)\pm O(\epsilon)}|z\rangle}{2^N}\left[\frac{|0\rangle - |1\rangle}{\sqrt{2}}\right]. \tag{4.6.14}$$

We notice that

$$(-1)^{\pm O(\epsilon)}|z\rangle = (e^{\pm i\pi O(\epsilon)})|z\rangle \simeq |z\rangle \tag{4.6.15}$$

because $e^{\pm i\pi O(\epsilon)} \simeq 1$. Thus, we have

$$|\psi_3\rangle \simeq \pm\sum_z\sum_x \frac{(-1)^{x\cdot z+a\cdot x}|z\rangle}{2^N}\left[\frac{|0\rangle - |1\rangle}{\sqrt{2}}\right]. \tag{4.6.16}$$

In what follows, we evaluate $\sum_x(-1)^{x\cdot z+a\cdot x}$. We notice that

$$\begin{aligned}
&\sum_x (e^{i\pi})^{x_1z_1+\cdots+x_Nz_N}(e^{i\pi})^{x_1a_1+\cdots+x_Na_N}\\
&= \sum_x (e^{i\pi})^{x_1(z_1+a_1)+\cdots+x_N(z_N+a_N)}\\
&= \sum_{x_1}(e^{i\pi})^{x_1(z_1+a_1)}\cdots\sum_{x_N}(e^{i\pi})^{x_N(z_N+a_N)}.
\end{aligned} \tag{4.6.17}$$

We have

$$\sum_{x_1}(e^{i\pi})^{x_1(z_1+a_1)} = (1 + (e^{i\pi z_1})(e^{i\pi a_1})). \tag{4.6.18}$$

By examining the cases $z_1 = 0$ and $z_1 = 1$ separately, we see that

$$\begin{aligned}
&(1 + e^{i\pi a_1})|0\rangle_1 + (1 - e^{i\pi a_1})|1\rangle_1 \\
&= 2e^{i\pi a_1/2}\frac{(e^{-i\pi a_1/2} + e^{i\pi a_1/2})}{2}|0\rangle_1 + 2ie^{i\pi a_1/2}\frac{(e^{-i\pi a_1/2} - e^{i\pi a_1/2})}{2i}|1\rangle_1 \\
&= 2(i)^{a_1}\cos(a_1\pi/2)|0\rangle_1 - i(i)^{a_1}2\sin(a_1\pi/2)|1\rangle_1.
\end{aligned} \tag{4.6.19}$$

Thus, we have

$$\begin{aligned}
|\psi_3\rangle &\simeq \pm\sum_z\sum_x \frac{(-1)^{x\cdot z + a\cdot x}|z\rangle}{2^N}\left[\frac{|0\rangle - |1\rangle}{\sqrt{2}}\right] \\
&= \pm\sum_z \frac{(1 + (-1)^{z_1}e^{i\pi a_1})\cdots(1 + (-1)^{z_N}e^{i\pi a_N})|z\rangle}{2^N}\left[\frac{|0\rangle - |1\rangle}{\sqrt{2}}\right] \\
&= \pm\prod_{j=1\ldots N}\left[(i)^{a_j}\cos(a_j\pi/2)|0\rangle_j - i(i)^{a_j}\sin(a_j\pi/2)|1\rangle_j\right]\left[\frac{|0\rangle - |1\rangle}{\sqrt{2}}\right].
\end{aligned} \tag{4.6.20}$$

We now observe a quantum state $|100\cdots 1\rangle$ with high probability if

$$\begin{aligned}
&(a_1 = 1 \pm \epsilon), \quad |\cos(a_1\pi/2)|^2 \ll |\sin(a_1\pi/2)|^2, \\
&(a_2 = \pm\epsilon), \quad |\cos(a_2\pi/2)|^2 \gg |\sin(a_2\pi/2)|^2, \\
&(a_3 = \pm\epsilon), \quad |\cos(a_3\pi/2)|^2 \gg |\sin(a_3\pi/2)|^2, \\
&\quad\vdots \qquad\qquad\qquad \vdots \\
&(a_N = 1 \pm \epsilon), \quad |\cos(a_N\pi/2)|^2 \ll |\sin(a_N\pi/2)|^2.
\end{aligned} \tag{4.6.21}$$

Thus, we present the Bernstein–Vazirani algorithm in a noisy environment.

We now introduce a success probability of finding a_1. It is the probability of detecting $|1\rangle_1$ if $a_1 = 1 \pm \epsilon$. On the other hand, an error probability of finding a_1 is as follows: it is the probability of detecting $|0\rangle_1$ if $a_1 = 1 \pm \epsilon$. In what follows, we evaluate the success probability of the algorithm. It is the probability that we detect the desirable quantum states for all $a_1, a_2, \ldots, a_N$.

The error probability for a_1 is

$$|\cos(a_1\pi/2)|^2 = E_1. \tag{4.6.22}$$

The error probability for a_2 is

$$|\sin(a_2\pi/2)|^2 = E_2 \tag{4.6.23}$$

and so on. The success probability for a_1 is

$$|\sin(a_1\pi/2)|^2 = 1 - E_1. \tag{4.6.24}$$

The success probability for a_2 is

$$|\cos(a_2\pi/2)|^2 = 1 - E_2 \tag{4.6.25}$$

and so on. The success probability S for the algorithm is

$$S = (1 - E_1)(1 - E_2)\cdots(1 - E_N). \tag{4.6.26}$$

The algorithm that we discuss evaluates simultaneously $a_1, a_2, \ldots, a_N$ with the success probability S. Thus, we can know the noisy function $g(x)$ with the success probability S.

We show that the quantum algorithm in the presence of environmental noise is superior to its classical counterpart by a factor of 2^N.

Parts of this chapter have been reprinted from Nagata K and Nakamura T 2017 Quantum cryptography, quantum communication, and quantum computer in a noisy environment *Int. J. Theor. Phys.* **56** 2086

IOP Publishing

Simplified Quantum Computing with Applications

Koji Nagata, Do Ngoc Diep, Ahmed Farouk and Tadao Nakamura

Chapter 5

Quantum communication based on Simon's algorithm

This chapter provides quantum communication based on Simon's algorithm.

5.1 Review of Simon's algorithm

Suppose that

$$f\colon \{0, 1\}^N \to \{0, 1\}^N \tag{5.1.1}$$

is a function with an N-bit domain and an N-bit range. We assume the following case

$$\begin{aligned} &f(x) = f(x \oplus s) \quad \text{for all } x, \\ &x \oplus s = (x_1 \oplus s_1, x_2 \oplus s_2, \dots, x_N \oplus s_N). \end{aligned} \tag{5.1.2}$$

Simon's algorithm combines quantum superposition with a property of quantum mechanics known as interference.

Let us follow the quantum states through Simon's algorithm. The input state is

$$|\psi_0\rangle = |0\rangle^{\otimes N}|0\rangle. \tag{5.1.3}$$

After the componentwise Hadamard transformations on the first N-bit states, we have

$$|\psi_1\rangle = \sum_{x\in\{0,1\}^N} \frac{|x\rangle}{\sqrt{2^N}}|0\rangle. \tag{5.1.4}$$

Next, the function f is evaluated using

$$U_f|x, y\rangle = |x, y \oplus f(x)\rangle \tag{5.1.5}$$

in giving

$$|\psi_2\rangle = \sum_x \frac{|x\rangle}{\sqrt{2^N}}|f(x)\rangle. \tag{5.1.6}$$

We have

$$|\psi_3\rangle = \sum_x \frac{|x \oplus s\rangle}{\sqrt{2^N}}|f(x)\rangle \tag{5.1.7}$$

by using $f(x) = f(x \oplus s)$. Thus, we have

$$|\psi_4\rangle = \frac{1}{2}(|\psi_2\rangle + |\psi_3\rangle) = \sum_x \frac{|x\rangle + |x \oplus s\rangle}{\sqrt{2^{N+2}}}|f(x)\rangle. \tag{5.1.8}$$

In what follows, we derive the result of the Hadamard transformation of $|x\rangle + |x \oplus s\rangle$. We have the very useful equation

$$H^{\otimes N}|x\rangle = \frac{\sum_z (-1)^{x \cdot z}|z\rangle}{\sqrt{2^N}}. \tag{5.1.9}$$

We also have

$$H^{\otimes N}|x \oplus s\rangle = \frac{\sum_z (-1)^{z \cdot (x \oplus s)}|z\rangle}{\sqrt{2^N}}. \tag{5.1.10}$$

Thus, we have

$$\begin{aligned} H^{\otimes N}(|x\rangle + |x \oplus s\rangle) &= \frac{\sum_z (-1)^{x \cdot z + z \cdot (x \oplus s)}|z\rangle}{\sqrt{2^N}} \\ &= \frac{\sum_z [(-1)^{x \cdot z}(1 + (-1)^{z \cdot s})]|z\rangle}{\sqrt{2^N}}. \end{aligned} \tag{5.1.11}$$

Therefore, if Alice measures $|z\rangle$, then we have

$$z \cdot s = 0. \tag{5.1.12}$$

Thus, if Alice measures $|z_1\rangle$, $|z_2\rangle$, … , $|z_N\rangle$, she gets the s.

5.2 Quantum communication based on Simon's algorithm

First, Alice and Bob promise to use a multi-periodic function f such as $f(x) = f(x \oplus s)$ for all x. Alice does not know the period vector s, but Bob does. Alice's goal is of evaluating with certainty what s Bob chooses. Alice prepares a suitable $(N + 1)$-partite uncorrelated state, performs the componentwise Hadamard transformations on the state, and sends the output state to Bob. Bob then performs

Simon's algorithm and inputs the information about the s into the final state. Alice asks him what state it is $O(N)$ times. Alice measures the final state and she knows the s. If the s is learned by Alice, then Alice and Bob share N bits of information, by $O(N)$ communications with each other. In the classical case, Alice needs at least $O(2^N)$ communications with Bob to get the s.

- First, Alice prepares the qubits in (5.1.4) and sends the $N + 1$ qubits to Bob.
- Next, Bob picks up an N bits period vector s and applies U_f as in (5.1.5) to (5.1.6) thereby the state is evolving the $N + 1$ qubits to give (5.1.6). He then sends the N qubits to Alice.
- Finally, Alice applies the Hadamard transformation to each of the first N qubits, and performs $O(N)$ measurements. She learns the period vector s. Alice and Bob now share N bits of information.
- In the classical counterpart, Alice needs $O(2^N)$ communications with Bob to get the period vector s because the possible combination of N bits for s is 2^N.

We thus show that quantum communication is superior to classical communication by a factor $O(2^N/N)$ in the case of Simon's algorithm.

Parts of this chapter have been reprinted from Nagata K, Nakamura T, Geurdes H, Batle J, Abdalla S and Farouk A 2017 Quantum communication based on Simon's algorithm *Int. J. Emerg. Eng. Res. Technol.* **5** 28–31 (https://www.ijeert.org/v5-i8)

IOP Publishing

Simplified Quantum Computing with Applications

Koji Nagata, Do Ngoc Diep, Ahmed Farouk and Tadao Nakamura

Chapter 6

Expansion of Deutsch's algorithm

Deutsch's algorithm determines if the given function is constant or balanced. The function is called to be constant if $f(0) = f(1)$. The function is called to be balanced if $f(0) \neq f(1)$. In this chapter, we expand Deutsch's algorithm based on the method proposed by Nagata and Nakamura. Deutsch's algorithm expanded determines all the mappings of the given function. We can determine simultaneously the following mappings: $f(0) = ?$ and $f(1) = ?$.

6.1 Expansion of Deutsch's algorithm for determining all the mappings of a function

Our discussion is based on Nielsen and Chuang. Quantum superposition and quantum phase factor are fundamental features of many quantum algorithms. Both of them are necessary. They allow quantum computers to evaluate simultaneously the mappings of a function $f(x)$ for many different x. Suppose

$$f: \{0, 1\} \rightarrow \{0, 1\} \tag{6.1.1}$$

is a function with a one-bit domain and range. A convenient way of computing the function on a quantum computer is of considering a two-qubit quantum computer that starts with the state $|x, y\rangle$, where x and y are variables used in mapping f. The abbreviation $|x, y\rangle$ stands for $|x\rangle \otimes |y\rangle$.

It is possible to transform the state $|x, y\rangle$ into

$$|x, y \oplus f(x)\rangle, \tag{6.1.2}$$

by applying the quantum oracle, where $\oplus$ indicates addition modulo 2. We denote the transformation U_f defined by the map

$$U_f|x, y\rangle = |x, y \oplus f(x)\rangle. \tag{6.1.3}$$

6.2 Deutsch's algorithm

Let us review Deutsch's formula as follows:

$$U_f|0\rangle(|0\rangle - |1\rangle)/\sqrt{2} = |0\rangle(|0 \oplus f(0)\rangle - |1 \oplus f(0)\rangle)/\sqrt{2}$$
$$= \begin{cases} (-1)^{f(0)}|0\rangle(|0\rangle - |1\rangle)/\sqrt{2} & \text{if } f(0) = 0, \\ (-1)^{f(0)}|0\rangle(|0\rangle - |1\rangle)/\sqrt{2} & \text{if } f(0) = 1. \end{cases} \tag{6.2.1}$$

$$U_f|1\rangle(|0\rangle - |1\rangle)/\sqrt{2} = |1\rangle(|0 \oplus f(1)\rangle - |1 \oplus f(1)\rangle)/\sqrt{2}$$
$$= \begin{cases} (-1)^{f(1)}|1\rangle(|0\rangle - |1\rangle)/\sqrt{2} & \text{if } f(1) = 0, \\ (-1)^{f(1)}|1\rangle(|0\rangle - |1\rangle)/\sqrt{2} & \text{if } f(1) = 1. \end{cases} \tag{6.2.2}$$

This is the phase kickback formation. We see $f(0)$ and $f(1)$ are in the superposition, i.e., the number of the mappings is two.

Generally speaking, so long as using $\left[\frac{|0\rangle - |1\rangle}{\sqrt{2}}\right]$ as its eigenstate for a gate, $|x\rangle$ of first qubit is changed into the $(-1)^{f(x)}|x\rangle$ that has the phase kickback factor $(-1)^{f(x)}$ after operation by the gate.

Let us introduce the Bloch sphere. We consider a quantum state lying in x-axis and a quantum state lying in z-axis. Deutsch's formula does not use a quantum state lying in y-axis. $f(0)$ and $f(1)$ appear in the phase factor, but we cannot obtain both of them at the same time. The phase of the result of (6.2.1) is the same as the phase of the result of (6.2.2), therefore, we cannot distinguish between the two.

In fact, we may define the input state as follows:

$$|\psi_0\rangle_d = \alpha|0\rangle|-\rangle_x + \beta|1\rangle|-\rangle_x,$$
$$_d\langle\psi_0|\psi_0\rangle_d = 1 \Leftarrow |\alpha|^2 + |\beta|^2 = 1, \alpha \neq 0, \beta \neq 0. \tag{6.2.3}$$

Applying U_{f_i}, $(i = 0, 1, 2, 3)$ to $|\psi_0\rangle_d$, $U_{f_i}|\psi_0\rangle_d = |\psi_1\rangle_{id}$, therefore leaves us with one of 2^2 cases:

$$\begin{aligned} U_{f_0}|\psi_0\rangle_d &= |\psi_1\rangle_{0d} = \alpha|0\rangle|-\rangle_x + \beta|1\rangle|-\rangle_x \quad \text{iff } f_0(0) = 0, f_0(1) = 0, \\ U_{f_1}|\psi_0\rangle_d &= |\psi_1\rangle_{1d} = -\alpha|0\rangle|-\rangle_x - \beta|1\rangle|-\rangle_x \quad \text{iff } f_1(0) = 1, f_1(1) = 1, \\ U_{f_2}|\psi_0\rangle_d &= |\psi_1\rangle_{2d} = \alpha|0\rangle|-\rangle_x - \beta|1\rangle|-\rangle_x \quad \text{iff } f_2(0) = 0, f_2(1) = 1, \\ U_{f_3}|\psi_0\rangle_d &= |\psi_1\rangle_{3d} = -\alpha|0\rangle|-\rangle_x + \beta|1\rangle|-\rangle_x \quad \text{iff } f_3(0) = 1, f_3(1) = 0 \end{aligned} \tag{6.2.4}$$

.

Thus, if we have (6.2.4), we do not obtain simultaneously both $f(0)$ and $f(1)$ by measuring the single output state.

By measuring $|\psi_1\rangle_{id}$, we cannot determine simultaneously all the two mappings of $f_i(x)$ for all x. But, we can determine if the given function is constant or balanced. This is very interesting indeed: the quantum algorithm gives us the ability to determine a property of $f_i(x)$. This is faster than a classical apparatus which would require at least two evaluations.

6.3 Expansion of Deutsch's algorithm

We can expand Deutsch's algorithm using a quantum state lying in xy-plane. In what follows, we consider the Bloch sphere, especially, we consider a quantum state lying in xy-plane. We expand the quantum state from the eigenstate. From Deutsch's formula and the mapping U_f, we arrive the following formulas:

$$U_f|0\rangle(\cos\frac{\theta}{2}|0\rangle + e^{i\phi}\sin\frac{\theta}{2}|1\rangle) = |0\rangle(\cos\frac{\theta}{2}|0\oplus f(0)\rangle + e^{i\phi}\sin\frac{\theta}{2}|1\oplus f(0)\rangle)$$
$$= \begin{cases} |0\rangle(\cos\frac{\theta}{2}|0\rangle + e^{i\phi}\sin\frac{\theta}{2}|1\rangle) & \text{if } f(0) = 0, \\ |0\rangle(\cos\frac{\theta}{2}|1\rangle + e^{i\phi}\sin\frac{\theta}{2}|0\rangle) & \text{if } f(0) = 1. \end{cases} \tag{6.3.1}$$

$$U_f|1\rangle(\cos\frac{\theta'}{2}|0\rangle + e^{i\phi'}\sin\frac{\theta'}{2}|1\rangle) = |1\rangle(\cos\frac{\theta'}{2}|0\oplus f(1)\rangle + e^{i\phi'}\sin\frac{\theta'}{2}|1\oplus f(1)\rangle)$$
$$= \begin{cases} |1\rangle(\cos\frac{\theta'}{2}|0\rangle + e^{i\phi'}\sin\frac{\theta'}{2}|1\rangle) & \text{if } f(1) = 0, \\ |1\rangle(\cos\frac{\theta'}{2}|1\rangle + e^{i\phi'}\sin\frac{\theta'}{2}|0\rangle) & \text{if } f(1) = 1. \end{cases} \tag{6.3.2}$$

This is enough to realize our main goal, but, to simplify, we suppose a quantum state lying in just y-axis and a quantum state lying in just x-axis. So let (θ, ϕ) be $(\pi/2, 3\pi/2)$ and let (θ', ϕ') be $(\pi/2, \pi)$ in giving

$$U_f|0\rangle(|0\rangle - i|1\rangle)/\sqrt{2} = \begin{cases} (-i)^{f(0)}|0\rangle(|0\rangle - i|1\rangle)/\sqrt{2} & \text{if } f(0) = 0, \\ (-i)^{f(0)}|0\rangle(|0\rangle + i|1\rangle)/\sqrt{2} & \text{if } f(0) = 1. \end{cases} \tag{6.3.3}$$

$$U_f|1\rangle(|0\rangle - |1\rangle)/\sqrt{2} = \begin{cases} (-1)^{f(1)}|1\rangle(|0\rangle - |1\rangle)/\sqrt{2} & \text{if } f(1) = 0, \\ (-1)^{f(1)}|1\rangle(|0\rangle - |1\rangle)/\sqrt{2} & \text{if } f(1) = 1. \end{cases} \tag{6.3.4}$$

The phase of the result of (6.3.3) is different from the phase of the result of (6.3.4). Let us take a summation, that is, (6.3.3) plus (6.3.4). Then we have (6.3.5). We note here that the first term is orthogonal to the second term. We expand the input state from a one-dimensional state to a two-dimensional state. There may be a phase factor in the two-dimensional state. A vector turns in the two-dimensional unitary space. We emphasize the fact is most important point in our quantum algorithm. $f(0)$ and $f(1)$ appear in the phase factor, and we can obtain both of them at the same time as we show below.

We define the input state as (6.3.5). Here we use a phase effect, which is a quantum phenomenon. We define the input state as follows, using an imaginary number i:

$$\begin{aligned} |\psi_0\rangle &= \alpha|0\rangle|-\rangle_y + \beta|1\rangle|-\rangle_x, \\ \langle\psi_0|\psi_0\rangle &= 1 \Leftarrow |\alpha|^2 + |\beta|^2 = 1, \alpha \neq 0, \beta \neq 0. \end{aligned} \tag{6.3.5}$$

Applying U_{f_i}, $(i = 0, 1, 2, 3)$ to $|\psi_0\rangle$, $U_{f_i}|\psi_0\rangle = |\psi_1\rangle_i$, therefore leaves us with one of 2^2 cases:

$$\begin{aligned}
U_{f_0}|\psi_0\rangle &= |\psi_1\rangle_0 = \alpha|0\rangle|-\rangle_y + \beta|1\rangle|-\rangle_x \quad \text{iff } f_0(0) = 0, f_0(1) = 0, \\
U_{f_1}|\psi_0\rangle &= |\psi_1\rangle_1 = -i\alpha|0\rangle|+\rangle_y - \beta|1\rangle|-\rangle_x \quad \text{iff } f_1(0) = 1, f_1(1) = 1, \\
U_{f_2}|\psi_0\rangle &= |\psi_1\rangle_2 = \alpha|0\rangle|-\rangle_y - \beta|1\rangle|-\rangle_x \quad \text{iff } f_2(0) = 0, f_2(1) = 1, \\
U_{f_3}|\psi_0\rangle &= |\psi_1\rangle_3 = -i\alpha|0\rangle|+\rangle_y + \beta|1\rangle|-\rangle_x \quad \text{iff } f_3(0) = 1, f_3(1) = 0.
\end{aligned} \tag{6.3.6}$$

If we have (6.3.6), we know simultaneously both $f(0)$ and $f(1)$ by measuring the single output state.

So, by measuring $|\psi_1\rangle_i$, we may determine simultaneously all the two mappings of $f_i(x)$ for all x. This is very interesting indeed: the quantum algorithm gives us the ability to determine a perfect property of $f_i(x)$, namely, $f_i(x)$ itself. This is faster than a classical apparatus which would require 2^2 evaluations.

Our algorithm is as follows:

- Select a function f_i and do not know any mappings of it, that is,

$$f_i(0) = ?, f_i(1) = ?. \tag{6.3.7}$$

- Operate U_{f_i} to $|\psi_0\rangle$ in giving $|\psi_1\rangle_i$.
- From $|\psi_1\rangle_i$, obtain the values of all the mappings concerning the function f_i.

It is very interesting to consider the Deutsch–Jozsa algorithm expanded which uses N qubits, and in this case, we could evaluate the values of the 2^N mappings at the same time.

Parts of this chapter have been reprinted from Nagata K and Nakamura T 2020 Generalization of Deutsch's algorithm *Int. J. Theor. Phys.* **59** 2557

IOP Publishing

Simplified Quantum Computing with Applications

Koji Nagata, Do Ngoc Diep, Ahmed Farouk and Tadao Nakamura

Chapter 7

Some theoretically organized algorithm for quantum computers

In this chapter, we first propose a novel parallel computation, still including today's algorithm methodology for quantum computing, for all of the combinations of values in variables of a logical function. Our concern so far has been to obtain an attribute of some function. In fact, such a task is only for one-task problem solving. However, we could treat positively the plural evaluations of some logical function in parallel instead of testing the function for finding out its attribute. In fact, these evaluations of the function are naturally included and evaluated, in parallel, in normal quantum computing discussing a function in a Boolean algebra stemmed from atoms in it. As is naturally understandable with mathematics, quantum computing naturally meets the category of a Boolean algebra. The reason why we positively introduce a Boolean algebra here is because we have multiple evaluations of a function in quantum computing general.

7.1 New type of quantum algorithm for determining the 2^1 mappings of a function

Quantum superposition is a fundamental feature of many quantum algorithms. It allows quantum computers to evaluate simultaneously the mappings of a function $f(x)$ for many different x. Suppose

$$f: \{0, 1\} \rightarrow \{0, 1\} \tag{7.1.1}$$

is a function with a one-bit domain and range. A convenient way of computing the function on a quantum computer is to consider a two-qubit quantum computer that starts with the state $|x, y\rangle$, where x and y are variables used in mapping f. The abbreviation $|x, y\rangle$ stands for $|x\rangle \otimes |y\rangle$.

Like the Deutsch–Jozsa problem, we are given a black box quantum computer known as an oracle that implements some function $f: \{0, 1\}^2 \rightarrow \{0, 1\}$. For the

doi:10.1088/978-0-7503-4700-6ch7

quantum algorithms to work, the oracle computing $f(x)$ from x has to be a quantum oracle which does not decohere x. It also must not leave any copy of x lying around at the end of the oracle call. We have the function f implemented as a quantum oracle. The oracle maps the state $|x\rangle \otimes |y\rangle$ to $|x\rangle \otimes |y \oplus f(x)\rangle$, where $\oplus$ is addition modulo 2.

It is possible to transform the state $|x, y\rangle$ into

$$|x, y \oplus f(x)\rangle, \tag{7.1.2}$$

by applying the quantum oracle, where $\oplus$ indicates addition modulo 2. We denote the transformation U_f defined by the mapping

$$U_f|x, y\rangle = |x, y \oplus f(x)\rangle. \tag{7.1.3}$$

Here (7.1.2) and (7.1.3) meet the category of a Boolean algebra. Later we see the fact from the results of section 7.2 using (7.1.2) and (7.1.3). Namely, the results meet the category of a Boolean algebra. Quantum computing meets the category of a Boolean algebra.

We want to develop the algorithms of quantum computers toward the ultimate parallel processing on them. The way to do this is to find out the very true ultimate parallelism, thinking of the physical quantum phenomena. From the usual phase kickback formation and the mapping U_f, we insert an imaginary number i and we can define the following formulas:

$$\begin{aligned} &U_f|0\rangle(|0\rangle - i|1\rangle)/\sqrt{2} = +|0\rangle(|f(0)\rangle - i|\overline{f(0)}\rangle)/\sqrt{2} \\ &= \begin{cases} (-i)^{f(0)}|0\rangle(|0\rangle - i|1\rangle)/\sqrt{2} & \text{if } f(0) = 0, \\ (-i)^{f(0)}|0\rangle(|0\rangle + i|1\rangle)/\sqrt{2} & \text{if } f(0) = 1. \end{cases} \end{aligned} \tag{7.1.4}$$

$$\begin{aligned} &U_f|1\rangle(|0\rangle - |1\rangle)/\sqrt{2} = +|1\rangle(|f(1)\rangle - |\overline{f(1)}\rangle)/\sqrt{2} \\ &= \begin{cases} (-1)^{f(1)}|1\rangle(|0\rangle - |1\rangle)/\sqrt{2} & \text{if } f(1) = 0, \\ (-1)^{f(1)}|1\rangle(|0\rangle - |1\rangle)/\sqrt{2} & \text{if } f(1) = 1, \end{cases} \end{aligned} \tag{7.1.5}$$

where $|\bar{1}\rangle = |0\rangle$ and $|\bar{0}\rangle = |1\rangle$. We use a combination between a unitary transformation theory and logic theory. In other words, we use a combination between a Pauli operator σ_x and NOT operation.

The phase of the result of (7.1.4) is different from the phase of the result of (7.1.5). Let us take a summation, that is, (7.1.4) plus (7.1.5). Then we have (7.1.7). We can solve a mathematical problem if we define the input state as (7.1.7) because we define the mapping U_f. Here we use a phase effect, which is a quantum phenomenon.

We define the following notations:

$$|-\rangle_y = \frac{|0\rangle - i|1\rangle}{\sqrt{2}}, \; |+\rangle_y = \frac{|0\rangle + i|1\rangle}{\sqrt{2}}, \; |-\rangle_x = \frac{|0\rangle - |1\rangle}{\sqrt{2}}. \tag{7.1.6}$$

We define the input state as follows, using an imaginary number i:

$$|\psi_0\rangle = \alpha|0\rangle|-\rangle_y + \beta|1\rangle|-\rangle_x, \langle\psi_0|\psi_0\rangle = 1 \Leftarrow |\alpha|^2 + |\beta|^2 = 1, \alpha \neq 0, \beta \neq 0. \quad (7.1.7)$$

Applying U_{f_i}, $(i = 0, 1, 2, 2^{2^1} - 1)$ to $|\psi_0\rangle$, $U_{f_i}|\psi_0\rangle = |\psi_1\rangle_i$, therefore leaves us with one of 2^{2^1} cases, where the power 1 of 2^{2^1} means the case of one qubit:

$$\begin{aligned}
|\psi_1\rangle_0 &= \alpha|0\rangle|-\rangle_y + \beta|1\rangle|-\rangle_x \quad \text{then } f_0(0) = 0, f_0(1) = 0,\\
|\psi_1\rangle_1 &= -i\alpha|0\rangle| + \rangle_y - \beta|1\rangle| - \rangle_x \quad \text{then } f_1(0) = 1, f_1(1) = 1,\\
|\psi_1\rangle_2 &= \alpha|0\rangle|-\rangle_y - \beta|1\rangle|-\rangle_x \quad \text{then } f_2(0) = 0, f_2(1) = 1,\\
|\psi_1\rangle_3 &= -i\alpha|0\rangle| + \rangle_y + \beta|1\rangle| - \rangle_x \quad \text{then } f_3(0) = 1, f_3(1) = 0,
\end{aligned} \quad (7.1.8)$$

where these equations have a property that the relation between each equation and the condition after 'then' is regarded as a 'if and only if' condition since we herein process all of the operations only under a unitary transformation. So, the conditions after 'then' are regarded as the results.

If we have (7.1.8), we know simultaneously both $f(0)$ and $f(1)$ by measuring the single output state. How do we have (7.1.8)? Note that we cannot solve it by using only the usual phase kickback formation. It changes only the global phase and we cannot distinguish between them. We avoid this situation.

So, by measuring $|\psi_1\rangle_i$, we may determine simultaneously all the 2^1 mappings of $f_i(x)$ for all x. This is very interesting indeed: the quantum algorithm gives us the ability to determine a perfect property of $f_i(x)$, namely, $f_i(x)$ itself. This is faster than a classical apparatus which would require 2^{2^1} evaluations.

Our algorithm is as follows:

- Select a function f_i under the assumption that we do not know the mappings concerning the function f_i, that is, $f_i(0) = ?, f_i(1) = ?$.
- Operate U_{f_i} to $|\psi_0\rangle$ in giving $|\psi_1\rangle_i$.
- From $|\psi_1\rangle_i$, obtain all the mappings concerning the function f_i.

7.2 New type of quantum algorithm for determining the 2^2 mappings of a function

We propose a quantum algorithm for determining the 2^2 mappings of a function. Suppose newly

$$f: \{0, 1\}^2 \to \{0, 1\} \quad (7.2.1)$$

is a function. We want to know simultaneously the 2^2 mappings $f(0, 0), f(0, 1), f(1, 0)$, and $f(1, 1)$. Later we see a complete matching between our results and a Boolean algebra F_2. In the Boolean algebra F_2, the function is a two-valuable function. For example, $f(x, y)$ is the function where x and y are variables used in mapping f. In what follows, the abbreviation $f(xy)$ stands for $f(x, y)$. We see a combination between a unitary transformation theory and logic theory.

We define the input state as follows, using an application of (7.1.7):

$$\begin{aligned} &|\psi_0\rangle = a_1|00\rangle|-\rangle_y + a_2|01\rangle|-\rangle_y + a_3|10\rangle|-\rangle_x + a_4|11\rangle|-\rangle_x, \\ &\langle\psi_0|\psi_0\rangle = 1 \Leftarrow |a_1|^2 + |a_2|^2 + |a_3|^2 + |a_4|^2 = 1, \\ &a_1 \neq 0,\ a_2 \neq 0,\ a_3 \neq 0,\ a_4 \neq 0. \end{aligned} \tag{7.2.2}$$

From the mapping U_f, we can define the following formulas:

$$U_f|00\rangle|-\rangle_y = \begin{cases} (-i)^{f(00)}|00\rangle|-\rangle_y & \text{if } f(00) = 0, \\ (-i)^{f(00)}|00\rangle|+\rangle_y & \text{if } f(00) = 1. \end{cases} \tag{7.2.3}$$

$$U_f|01\rangle|-\rangle_y = \begin{cases} (-i)^{f(01)}|01\rangle|-\rangle_y & \text{if } f(01) = 0, \\ (-i)^{f(01)}|01\rangle|+\rangle_y & \text{if } f(01) = 1. \end{cases} \tag{7.2.4}$$

$$U_f|10\rangle|-\rangle_x = \begin{cases} (-1)^{f(10)}|10\rangle|-\rangle_x & \text{if } f(10) = 0, \\ (-1)^{f(10)}|10\rangle|-\rangle_x & \text{if } f(10) = 1. \end{cases} \tag{7.2.5}$$

$$U_f|11\rangle|-\rangle_x = \begin{cases} (-1)^{f(11)}|11\rangle|-\rangle_x & \text{if } f(11) = 0, \\ (-1)^{f(11)}|11\rangle|-\rangle_x & \text{if } f(11) = 1. \end{cases} \tag{7.2.6}$$

Applying U_{f_i}, $(i = 0, 1, 2, \ldots, 2^{2^2} - 1)$ to $|\psi_0\rangle$, $U_{f_i}|\psi_0\rangle = |\psi_1\rangle_i$, therefore leaves us with one of 2^{2^2} cases:

$$\begin{aligned} &|\psi_1\rangle_0 = a_1|00\rangle|-\rangle_y + a_2|01\rangle|-\rangle_y + a_3|10\rangle|-\rangle_x + a_4|11\rangle|-\rangle_x \\ &\text{iff } f_0(00) = 0, f_0(01) = 0, f_0(10) = 0, f_0(11) = 0, \end{aligned} \tag{7.2.7}$$

$$\begin{aligned} &|\psi_1\rangle_1 = a_1|00\rangle|-\rangle_y + a_2|01\rangle|-\rangle_y + a_3|10\rangle|-\rangle_x - a_4|11\rangle|-\rangle_x \\ &\text{iff } f_1(00) = 0, f_1(01) = 0, f_1(10) = 0, f_1(11) = 1, \end{aligned} \tag{7.2.8}$$

$$\begin{aligned} &|\psi_1\rangle_2 = a_1|00\rangle|-\rangle_y + a_2|01\rangle|-\rangle_y - a_3|10\rangle|-\rangle_x + a_4|11\rangle|-\rangle_x \\ &\text{iff } f_2(00) = 0, f_2(01) = 0, f_2(10) = 1, f_2(11) = 0, \end{aligned} \tag{7.2.9}$$

$$\begin{aligned} &|\psi_1\rangle_3 = a_1|00\rangle|-\rangle_y + a_2|01\rangle|-\rangle_y - a_3|10\rangle|-\rangle_x - a_4|11\rangle|-\rangle_x \\ &\text{iff } f_3(00) = 0, f_3(01) = 0, f_3(10) = 1, f_3(11) = 1, \end{aligned} \tag{7.2.10}$$

$$\begin{aligned} &|\psi_1\rangle_4 = a_1|00\rangle|-\rangle_y - ia_2|01\rangle|+\rangle_y + a_3|10\rangle|-\rangle_x + a_4|11\rangle|-\rangle_x \\ &\text{iff } f_4(00) = 0, f_4(01) = 1, f_4(10) = 0, f_4(11) = 0, \end{aligned} \tag{7.2.11}$$

$$\begin{aligned} &|\psi_1\rangle_5 = a_1|00\rangle|-\rangle_y - ia_2|01\rangle|+\rangle_y + a_3|10\rangle|-\rangle_x - a_4|11\rangle|-\rangle_x \\ &\text{iff } f_5(00) = 0, f_5(01) = 1, f_5(10) = 0, f_5(11) = 1, \end{aligned} \tag{7.2.12}$$

$$|\psi_1\rangle_6 = a_1|00\rangle|-\rangle_y - ia_2|01\rangle|+\rangle_y - a_3|10\rangle|-\rangle_x + a_4|11\rangle|-\rangle_x \\ \text{iff } f_6(00) = 0, f_6(01) = 1, f_6(10) = 1, f_6(11) = 0, \tag{7.2.13}$$

$$|\psi_1\rangle_7 = a_1|00\rangle|-\rangle_y - ia_2|01\rangle|+\rangle_y - a_3|10\rangle|-\rangle_x - a_4|11\rangle|-\rangle_x \\ \text{iff } f_7(00) = 0, f_7(01) = 1, f_7(10) = 1, f_7(11) = 1, \tag{7.2.14}$$

$$|\psi_1\rangle_8 = -ia_1|00\rangle|+\rangle_y + a_2|01\rangle|-\rangle_y + a_3|10\rangle|-\rangle_x + a_4|11\rangle|-\rangle_x \\ \text{iff } f_8(00) = 1, f_8(01) = 0, f_8(10) = 0, f_8(11) = 0, \tag{7.2.15}$$

$$|\psi_1\rangle_9 = -ia_1|00\rangle|+\rangle_y + a_2|01\rangle|-\rangle_y + a_3|10\rangle|-\rangle_x - a_4|11\rangle|-\rangle_x \\ \text{iff } f_9(00) = 1, f_9(01) = 0, f_9(10) = 0, f_9(11) = 1, \tag{7.2.16}$$

$$|\psi_1\rangle_{10} = -ia_1|00\rangle|+\rangle_y + a_2|01\rangle|-\rangle_y - a_3|10\rangle|-\rangle_x + a_4|11\rangle|-\rangle_x \\ \text{iff } f_{10}(00) = 1, f_{10}(01) = 0, f_{10}(10) = 1, f_{10}(11) = 0, \tag{7.2.17}$$

$$|\psi_1\rangle_{11} = -ia_1|00\rangle|+\rangle_y + a_2|01\rangle|-\rangle_y - a_3|10\rangle|-\rangle_x - a_4|11\rangle|-\rangle_x \\ \text{iff } f_{11}(00) = 1, f_{11}(01) = 0, f_{11}(10) = 1, f_{11}(11) = 1, \tag{7.2.18}$$

$$|\psi_1\rangle_{12} = -ia_1|00\rangle|+\rangle_y - ia_2|01\rangle|+\rangle_y + a_3|10\rangle|-\rangle_x + a_4|11\rangle|-\rangle_x \\ \text{iff } f_{12}(00) = 1, f_{12}(01) = 1, f_{12}(10) = 0, f_{12}(11) = 0, \tag{7.2.19}$$

$$|\psi_1\rangle_{13} = -ia_1|00\rangle|+\rangle_y - ia_2|01\rangle|+\rangle_y + a_3|10\rangle|-\rangle_x - a_4|11\rangle|-\rangle_x \\ \text{iff } f_{13}(00) = 1, f_{13}(01) = 1, f_{13}(10) = 0, f_{13}(11) = 1, \tag{7.2.20}$$

$$|\psi_1\rangle_{14} = -ia_1|00\rangle|+\rangle_y - ia_2|01\rangle|+\rangle_y - a_3|10\rangle|-\rangle_x + a_4|11\rangle|-\rangle_x \\ \text{iff } f_{14}(00) = 1, f_{14}(01) = 1, f_{14}(10) = 1, f_{14}(11) = 0, \tag{7.2.21}$$

$$|\psi_1\rangle_{15} = -ia_1|00\rangle|+\rangle_y - ia_2|01\rangle|+\rangle_y - a_3|10\rangle|-\rangle_x - a_4|11\rangle|-\rangle_x \\ \text{iff } f_{15}(00) = 1, f_{15}(01) = 1, f_{15}(10) = 1, f_{15}(11) = 1, \tag{7.2.22}$$

So, by measuring $|\psi_1\rangle_i$, we may determine simultaneously all the 2^2 mappings of $f_i(x, y)$ for all x and y. This is very interesting indeed: the quantum algorithm gives us the ability to determine a perfect property of $f_i(x, y)$, namely, $f_i(x, y)$ itself. This is faster than a classical apparatus which would require 2^{2^2} evaluations.

For efficiency of operations in quantum oracle computing to the classical, we must note how many functions are used in an algorithm. Here in equations (7.2.7)–(7.2.22), we see that the efficiency of this quantum operation seems to be 2^2 times the operation speed in classical operations if we limit our consideration within each function $|\psi_1\rangle_i$. However, thinking of the Bernstein–Vazirani algorithm-like quantum computing, we could see the efficiency appears to be 2^{2^2} times the operation speed of the classical. Briefly speaking, the efficiency depends upon the comparative object whenever we compare the two, which fact looks to be in our evaluations on comparison general.

Later we discuss the relation between set theory based upon atoms and our results in terms of a Boolean algebra. In particular, the result reveals a complete matching between quantum computing and a Boolean algebra. As is naturally understandable with mathematics, quantum computing belongs to the category of a Boolean algebra. We positively mention that the fundamental structures of quantum computing and von Neumann architecture are the same in terms of the category of a Boolean algebra. However, the main difference is based on parallelism for determining all the mappings used, especially in quantum computing.

7.3 Example using a logical function

Let us consider the case where $i = 1$. The logical function is as follows:

$$f_1(x, y) = A \wedge B. \tag{7.3.1}$$

where x and y are variables used in mapping f. $x(=0, 1)$ is variable for A. $y(=0, 1)$ is variable for B. We want to know simultaneously all the mappings:

$$f_1(0, 0), f_1(0, 1), f_1(1, 0), f_1(1, 1). \tag{7.3.2}$$

In the classical case we require 2^{2^2} evaluations. In the quantum case we require just one query.

The input state is as follows:

$$|\psi_0\rangle = a_1|00\rangle|-\rangle_y + a_2|01\rangle|-\rangle_y + a_3|10\rangle|-\rangle_x + a_4|11\rangle|-\rangle_x. \tag{7.3.3}$$

Applying U_{f_1} to $|\psi_0\rangle$, $U_{f_1}|\psi_0\rangle = |\psi_1\rangle_1$, we have the following output state:

$$|\psi_1\rangle_1 = a_1|00\rangle|-\rangle_y + a_2|01\rangle|-\rangle_y + a_3|10\rangle|-\rangle_x - a_4|11\rangle|-\rangle_x. \tag{7.3.4}$$

Therefore, we obtain simultaneously all the mappings of $f_1(x, y)$:

$$f_1(0, 0) = 0, f_1(0, 1) = 0, f_1(1, 0) = 0, f_1(1, 1) = 1. \tag{7.3.5}$$

This is faster than a classical apparatus which would require at least 2^{2^2} evaluations. Likewise, we can evaluate the 2^{2^2} functions in a Boolean algebra F_2.

7.4 New type of quantum algorithm for determining the 2^N mappings of a function

We propose a quantum algorithm for determining the 2^N mappings of a function. N means the number of qubits for our algorithm.

Quantum superposition is a fundamental feature of many quantum algorithms. It allows quantum computers to evaluate simultaneously the mappings of a function $f(x_1, x_2, \ldots, x_N)$ for many different $x_1, x_2, \ldots, x_N$. Suppose newly

$$f: \{0, 1\}^N \rightarrow \{0, 1\} \tag{7.4.1}$$

is a function.

We define the input state as follows, using an application of (7.1.7):

$$
\begin{aligned}
|\psi_0\rangle &= \sum_{j=0}^{2^{(N-1)}-1} a_j|j\rangle\left[\frac{|0\rangle - i|1\rangle}{\sqrt{2}}\right] + \sum_{k=2^{(N-1)}}^{2^N-1} a_k|k\rangle\left[\frac{|0\rangle - |1\rangle}{\sqrt{2}}\right], \\
\langle\psi_0|\psi_0\rangle &= 1 \Leftarrow |a_0|^2 + |a_1|^2 + \cdots + |a_{2^N-1}|^2 = 1, \\
a_0 &\neq 0,\ a_1 \neq 0,\ \ldots,\ a_{2^N-1} \neq 0.
\end{aligned}
\tag{7.4.2}
$$

Applying U_{f_i}, $(i = 0, 1, 2, \ldots, 2^{2^N} - 1)$ to $|\psi_0\rangle$, $U_{f_i}|\psi_0\rangle = |\psi_1\rangle_i$, therefore leaves us with one of 2^{2^N} cases:

$$
\begin{aligned}
|\psi_1\rangle_i = &\sum_{j=0}^{2^{(N-1)}-1} (-i)^{f_i(j)} a_j|j\rangle\left[\frac{|0\rangle - (-1)^{f_i(j)} i|1\rangle}{\sqrt{2}}\right] \\
&+ \sum_{k=2^{(N-1)}}^{2^N-1} (-1)^{f_i(k)} a_k|k\rangle\left[\frac{|0\rangle - |1\rangle}{\sqrt{2}}\right].
\end{aligned}
\tag{7.4.3}
$$

So, by measuring $|\psi_1\rangle_i$, we may determine simultaneously all the 2^N mappings of $f_i(x_1, x_2, \ldots, x_N)$ for all $x_1, x_2, \ldots, x_N$. This is very interesting indeed: the quantum algorithm gives us the ability to determine a perfect property of $f_i(x_1, x_2, \ldots, x_N)$, namely, $f_i(x_1, x_2, \ldots, x_N)$ itself. This is faster than a classical apparatus which would require at least 2^{2^N} evaluations.

7.5 Relation between set-theoretic atoms and the result in section 7.2

Let us discuss the relation between set-theoretic atoms and the result in section 7.2. The A and B are subsets which are constructed using the atoms $f1$ through $f4$ that are disjoint to one another. For example, newly using fi as an element of a Boolean algebra F_2,

$$
\begin{aligned}
A &= f1 \vee f3 = \{f1, f3\}, \\
B &= f1 \vee f2 = \{f1, f2\},
\end{aligned}
\tag{7.5.1}
$$

where,

$$
\begin{aligned}
f1 &= A \wedge B, \\
f2 &= A' \wedge B, \\
f3 &= A \wedge B', \\
f4 &= A' \wedge B'.
\end{aligned}
\tag{7.5.2}
$$

We can introduce a Boolean algebra F_2 as a power set of the atoms. F_2 is based on the value '1' of the two-variable switching functions. An atom is a function including only one '1' as its mapped value, in the four combinations of the values of A and B for the two-variable function.

Clearly we notice a complete matching between the Boolean algebra F_2 and our result in section 7.2. In fact we can see that equations (7.2.8), (7.2.9), (7.2.11), and

(7.2.15) are regarded as the four atoms of the Boolean algebra F_2. For example, we notice (7.2.8) OR (7.2.9) is equal to (7.2.10) and all elements are derived from the four atoms.

We see that the relation between set theory based upon atoms and our result in terms of a Boolean algebra. The important point is that we obtain all the elements of F_2 by means of a power set of atoms when we get the four atoms. So we can say that next our aim is of getting simultaneously (7.2.8), (7.2.9), (7.2.11), and (7.2.15). That means we get simultaneously (7.2.7)–(7.2.22) (all 16 patterns!).

We positively stress that the fundamental structures of quantum computing and von Neumann architecture are the same in terms of the category of a Boolean algebra. However, the main difference is based on parallelism for determining all the mappings used especially in quantum computing. We hope our discussions conclude the very true ultimate importance of quantum parallelism to construct quantum computers beyond von Neumann architecture.

The key of this section is to develop the algorithms of quantum computers toward the ultimate parallel processing on them. The way to do this is to find out the very true ultimate parallelism, thinking of the physical quantum phenomena. The algorithm developed here is toward the full uses of the features of quantum computers. The algorithm implies the ability of such computation based upon the concept of a Boolean algebra. Finally we have the ultimate computation for today's quantum computers.

Parts of this chapter have been reprinted from Nagata K and Nakamura T 2020 Some theoretically organized algorithm for quantum computers *Int. J. Theor. Phys.* **59** 611

Chapter 8

Some multi-quantum computing on quantum gating computers beyond a von Neumann architecture

In this chapter, we demonstrate a parallel computation in evaluating two functions stemming from their atoms related to some variables of two logical functions. These evaluations of the functions are included and evaluated, in parallel, in normal quantum computing, discussing two functions in a Boolean algebra stemming from atoms in it. As is understandable with mathematics, quantum computing with qubit systems is included and exemplified by a Boolean algebra, which treats only both 0 and 1. Namely, the theory in this chapter is natural in a logical sense even in a physics domain. Therefore, quantum computing has an ability to solve some mathematical problems described in a Boolean algebra. The reason why we introduce a Boolean algebra is because we can have multiple evaluations of two functions in quantum computing general. We show the ultimate importance of quantum parallelism to construct quantum computers beyond von Neumann architecture.

8.1 Quantum algorithm for determining all the mappings of two logical functions

Two-level systems are included and exemplified by a Boolean algebra, which treats only both 0 and 1. Namely, the theory in this chapter is quite natural in a logical sense even in a physics domain. Therefore, quantum computing has an ability to solve some mathematical problems described in a Boolean algebra.

Let us consider two logical functions which are constructed based on a Boolean algebra stemming from atoms in it: there are the 2^{2^2} functions $(f_0, f_1, \ldots, f_{15})$ in a Boolean algebra F_2. Here, we consider f_1 and f_{12}:

$$f_1(x, y) = A \wedge B, f_{12}(x, y) = A', \tag{8.1.1}$$

doi:10.1088/978-0-7503-4700-6ch8

where x and y are variables used in mapping f. A' is the same as logical notation $\neg A$. $x(=0, 1)$ is the variable for A. $y(=0, 1)$ is the variable for B. We make a combination between a unitary transformation theory and logic theory. We want to know simultaneously all the $2 \times 2^2 = 2^3$ mappings:

$$\begin{aligned} &f_1(00) = ?, f_1(01) = ?, f_1(10) = ?, f_1(11) = ?, \\ &f_{12}(00) = ?, f_{12}(01) = ?, f_{12}(10) = ?, f_{12}(11) = ?. \end{aligned} \tag{8.1.2}$$

In the classical case we require $2^{2^3} = 256$ evaluations. In the quantum case we require just a query.

8.2 Overview of the quantum algorithm

Our quantum algorithm is of determining all the 2^2 mappings of the following logical function:

$$\frac{f_1(x, y) + if_{12}(x, y)}{\sqrt{2}}. \tag{8.2.1}$$

- Input state is defined as follows:

$$a_1|00\rangle| - \rangle_y + a_2|01\rangle| - \rangle_y + a_3|10\rangle| - \rangle_x + a_4|11\rangle| - \rangle_x. \tag{8.2.2}$$

- Output state is defined as follows:

$$\frac{|\psi_1\rangle_1 + i|\psi_1\rangle_{12}}{\sqrt{2}}. \tag{8.2.3}$$

It is very important that $|\psi_1\rangle_1$ and $|\psi_1\rangle_{12}$ are orthogonal. If they are not so, we cannot distinguish between them. By measuring $|\psi_1\rangle_1$, we may determine simultaneously all the 2^2 mappings of a logical function $f_1(x, y) = A \wedge B$ for all x and y:

$$f_1(00) = 0, f_1(01) = 0, f_1(10) = 0, f_1(11) = 1. \tag{8.2.4}$$

By measuring $|\psi_1\rangle_{12}$, we may determine simultaneously all the 2^2 mappings of a logical function $f_{12}(x, y) = A'$ for all x and y:

$$f_{12}(00) = 1, f_{12}(01) = 1, f_{12}(10) = 0, f_{12}(11) = 0. \tag{8.2.5}$$

We define the superposition of them, namely, $\frac{|\psi_1\rangle_1 + i|\psi_1\rangle_{12}}{\sqrt{2}}$. By measuring $\frac{|\psi_1\rangle_1 + i|\psi_1\rangle_{12}}{\sqrt{2}}$, we may determine all the 2^2 mappings of the following logical function:

$$\frac{f_1(x, y) + if_{12}(x, y)}{\sqrt{2}}. \tag{8.2.6}$$

From the real part and the imaginary part, we may determine simultaneously all the 2^3 mappings of two logical functions $f_1(x, y) = A \wedge B$ and $f_{12}(x, y) = A'$ for all x and y:

$$\begin{aligned} &f_1(00) = 0, f_1(01) = 0, f_1(10) = 0, f_1(11) = 1, \\ &f_{12}(00) = 1, f_{12}(01) = 1, f_{12}(10) = 0, f_{12}(11) = 0. \end{aligned} \tag{8.2.7}$$

This is very interesting indeed: the quantum algorithm gives us the ability to determine a perfect property of the two logical functions $f_1(x, y)$ and $f_{12}(x, y)$, namely, $f_1(x, y)$ and $f_{12}(x, y)$ themselves. This is faster than a classical apparatus which would require at least 2^{2^3} evaluations.

Our algorithm is as follows:

- Select the two of logical functions f_1, f_{12} which are constructed based on a Boolean algebra stemming from atoms in it.
- Operate $\frac{U_{f_1} + iU_{f_{12}}}{\sqrt{2}}$ to $|\psi_0\rangle$ in giving $\frac{|\psi_1\rangle_1 + i|\psi_1\rangle_{12}}{\sqrt{2}}$.
- From $\frac{|\psi_1\rangle_1 + i|\psi_1\rangle_{12}}{\sqrt{2}}$, obtain simultaneously all the mappings concerning the two logical functions f_1, f_{12}.

8.3 Orthogonal pairs

It is very important that $|\psi_1\rangle_1$ and $|\psi_1\rangle_{12}$ are orthogonal. If they are not so, we cannot distinguish between the two. There are other pairs that are orthogonal each other. We notice the following relations:

$$\begin{aligned} {}_0\langle\psi_1|\psi_1\rangle_{13} &= 0, \\ {}_1\langle\psi_1|\psi_1\rangle_{12} &= 0, \\ {}_2\langle\psi_1|\psi_1\rangle_{15} &= 0, \\ {}_3\langle\psi_1|\psi_1\rangle_{14} &= 0, \\ {}_4\langle\psi_1|\psi_1\rangle_{9} &= 0, \\ {}_5\langle\psi_1|\psi_1\rangle_{8} &= 0, \\ {}_6\langle\psi_1|\psi_1\rangle_{11} &= 0, \\ {}_7\langle\psi_1|\psi_1\rangle_{10} &= 0. \end{aligned} \tag{8.3.1}$$

Thus, we can construct the similar example using two logical functions f_0, f_{13} and so on.

8.4 Quantum algorithm for determining all the mappings of all 16 two-variable functions

Originally, computers, even if they are quantum computers, should be operated in parallel as much as possible. Fortunately, our challenge has much success in combining quantum computing with Boolean algebra. Using this trial, we furthermore challenge as many Boolean functions as possible to operate in parallel. Therefore, we want to evaluate all the 16 two-variable functions at the same time. So, we introduce a parallel computing method described as follows:

By the 16 parallel computings, we obtain the following quantum state and we can evaluate simultaneously all the 16 two-variable functions:

$$|\psi_1\rangle_0 \otimes |\psi_1\rangle_1 \otimes \cdots \otimes |\psi_1\rangle_{15}. \tag{8.4.1}$$

Clearly, we can construct 16 parallel computings. On the other hand, we can construct 8 parallel computings in order to evaluate all the 16 two-variable functions by using the algorithm described the previous sections as follows:

In fact, by the 8 parallel computings, we obtain the following quantum state and we can evaluate simultaneously all the 16 two-variable functions:

$$\begin{aligned} &\frac{|\psi_1\rangle_0 + i|\psi_1\rangle_{13}}{\sqrt{2}} \otimes \frac{|\psi_1\rangle_1 + i|\psi_1\rangle_{12}}{\sqrt{2}} \otimes \frac{|\psi_1\rangle_2 + i|\psi_1\rangle_{15}}{\sqrt{2}} \otimes \frac{|\psi_1\rangle_3 + i|\psi_1\rangle_{14}}{\sqrt{2}} \otimes \\ &\frac{|\psi_1\rangle_4 + i|\psi_1\rangle_9}{\sqrt{2}} \otimes \frac{|\psi_1\rangle_5 + i|\psi_1\rangle_8}{\sqrt{2}} \otimes \frac{|\psi_1\rangle_6 + i|\psi_1\rangle_{11}}{\sqrt{2}} \otimes \frac{|\psi_1\rangle_7 + i|\psi_1\rangle_{10}}{\sqrt{2}}. \end{aligned} \tag{8.4.2}$$

Thus, by 8 parallel computings, we evaluate simultaneously all the 16 two-variable functions.

We have tried to demonstrate some parallel evaluations of two two-variable functions which are elements of a Boolean algebra composed of the four-atom set. These eight evaluations of the function have mathematically been evaluated in parallel. The reason why we have introduced a Boolean algebra is because quantum gating computing with qubit systems is fundamentally using only both 0 and 1. Then quantum computing has the ability to show some mathematically parallel computing in a Boolean algebra. Further, we have confirmed that 'potentially' 16 two-variable functions are in parallel evaluated at the same time. Thus, we have confirmed that in quantum gating computers some algorithms can have an ability that well-structured algorithms based on Boolean algebras can have mathematically parallel evaluations beyond von Neumann architecture in terms of processing time.

In order to demonstrate the fact that a quantum gate method can reveal a very true parallelism used so far by von Neumann style computers, we challenge a new algorithm for evaluating two logical functions which are constructed based on a Boolean algebra stemming from atoms in it. As is understandable with mathematics, quantum computing with qubit systems is included and exemplified by a Boolean algebra, which treats only both 0 and 1. Namely, the theory in this chapter is natural in logical sense even in a physics domain. Therefore, quantum computing has an ability to solve some mathematical problems described in a Boolean algebra. The reason why we introduce a Boolean algebra is because we have simple multiple evaluations of two functions in quantum computing general. We hope the challenge here will become an initiative for the ultimate way to realize a very true parallelism in quantum computers beyond von Neumann architecture.

Parts of this chapter have been reprinted from Nagata K and Nakamura T 2020 Quantum algorithm for evaluating two of logical functions simultaneously *Int. J. Theor. Phys.* **59** 2191

IOP Publishing

Simplified Quantum Computing with Applications

Koji Nagata, Do Ngoc Diep, Ahmed Farouk and Tadao Nakamura

Chapter 9

Quantum cryptography based on an algorithm for determining simultaneously all the mappings of a logical function

In this chapter, we study quantum cryptography based on an algorithm for determining simultaneously all the mappings of a logical function using an entangled state. The security of our cryptography is based on the Ekert 1991 protocol, which uses an entangled state. Eavesdropping destroys the entanglement. Alice selects a secret function from the number of possible function types. Bob's aim is then to determine the selected function (a key) without an eavesdropper learning it. In order for both Alice and Bob to be able to select the same function classically, in the worst case Bob requires multiple queries to Alice. In the quantum case, however, Bob requires just a single query. By measuring the single entangled state, which is sent to him by Alice, Bob can obtain the function that Alice selected. This quantum key distribution method is faster compared to the multiple queries that would be required in the classical case.

9.1 Quantum algorithm for determining all the two mappings of a logical function

In this section, we propose a quantum cryptography based on an algorithm for determining a logical function using qubit systems. Such a function assuming values in the set $\{T, F\}$ is called a logical function. We consider a logical function f: $\{0, 1\} \rightarrow \{0, 1\}$. Alice knows all the two mappings $f(0)$ and $f(1)$ of the function, that is, $f(x)$ itself. Bob knows none of them. His aim is to obtain all of the mappings without an eavesdropper learning them. In the classical case, Bob needs 2^2 queries. In the quantum case, Bob needs just a single query. Hence, the quantum cryptography is faster than classical cryptography by a factor of 2^2.

doi:10.1088/978-0-7503-4700-6ch9

Quantum superposition is a fundamental feature of many quantum algorithms. It allows quantum computers to evaluate simultaneously the mappings of a function $f(x)$ for many different values of x. Suppose that

$$f\colon \{0, 1\} \to \{0, 1\} \tag{9.1.1}$$

is a logical function with a one-bit domain and range. A convenient way of computing the function on a quantum computer is to consider a two-qubit quantum computer that starts with the state $|x, y\rangle$, where x and y are variables used in mapping f. The abbreviation $|x, y\rangle$ stands for $|x\rangle \otimes |y\rangle$.

Like in the Deutsch–Jozsa problem, we are given a black box quantum computer known as an oracle that implements some function $f\colon \{0, 1\}^2 \to \{0, 1\}$. For the quantum algorithms to work, the oracle computing $f(x)$ from x has to be a quantum oracle that does not decohere x. It also must not leave any copy of x lying around at the end of an oracle call. We have the function f implemented as a quantum oracle. The oracle maps the state $|x\rangle \otimes |y\rangle$ to $|x\rangle \otimes |y \oplus f(x)\rangle$, where $\oplus$ stands for addition modulo 2.

It is possible to transform the state $|x, y\rangle$ into

$$|x, y \oplus f(x)\rangle, \tag{9.1.2}$$

by applying the quantum oracle. Let U_f denote the transformation defined by the mapping

$$U_f|x, y\rangle = |x, y \oplus f(x)\rangle. \tag{9.1.3}$$

Here, (9.1.2) and (9.1.3) meet the category of Boolean algebras because their outcomes meet this category. Therefore, quantum computing meets the category of Boolean algebras.

We want to develop quantum algorithms that would allow for the ultimate parallel processing. The way to do it is to find the actual ultimate parallelism while keeping in mind the physical quantum phenomena. To that end, we insert an imaginary number i into the usual phase kickback formation and the mapping U_f, and define the following formulas:

$$\begin{aligned} U_f|0\rangle(|0\rangle - i|1\rangle)/\sqrt{2} &= +|0\rangle(|f(0)\rangle - i|\bar{f}(0)\rangle)/\sqrt{2} \\ &= \begin{cases} (-i)^{f(0)}|0\rangle(|0\rangle - i|1\rangle)/\sqrt{2} & \text{if } f(0) = 0, \\ (-i)^{f(0)}|0\rangle(|0\rangle + i|1\rangle)/\sqrt{2} & \text{if } f(0) = 1. \end{cases} \end{aligned} \tag{9.1.4}$$

$$\begin{aligned} U_f|1\rangle(|0\rangle - |1\rangle)/\sqrt{2} &= +|1\rangle(|f(1)\rangle - |\bar{f}(1)\rangle)/\sqrt{2} \\ &= \begin{cases} (-1)^{f(1)}|1\rangle(|0\rangle - |1\rangle)/\sqrt{2} & \text{if } f(1) = 0, \\ (-1)^{f(1)}|1\rangle(|0\rangle - |1\rangle)/\sqrt{2} & \text{if } f(1) = 1, \end{cases} \end{aligned} \tag{9.1.5}$$

where $|\bar{1}\rangle = |0\rangle$ and $|\bar{0}\rangle = |1\rangle$.

The phase of the outcome of (9.1.4) is different from the phase of the outcome of (9.1.5). Adding (9.1.4) and (9.1.5) gives (9.1.7). A mathematical problem can be solved if the input state is defined as (9.1.7) because the mapping U_f is defined. Here we use a phase effect, which is a quantum phenomenon.

We define the following notations:

$$|-\rangle_y = \frac{|0\rangle - i|1\rangle}{\sqrt{2}},\ |+\rangle_y = \frac{|0\rangle + i|1\rangle}{\sqrt{2}},\ |-\rangle_x = \frac{|0\rangle - |1\rangle}{\sqrt{2}}. \tag{9.1.6}$$

We further define the input state as follows, using an imaginary number i:

$$|\psi_0\rangle = \alpha|0\rangle|-\rangle_y + \beta|1\rangle|-\rangle_x,\ \langle\psi_0|\psi_0\rangle = 1 \Leftarrow |\alpha|^2 + |\beta|^2 = 1,\ \alpha \neq 0,\ \beta \neq 0. \tag{9.1.7}$$

Applying U_{f_i}, $(i = 0, 1, 2, 2^{2^1} - 1)$, to $|\psi_0\rangle$, results in $U_{f_i}|\psi_0\rangle = |\psi_1\rangle_i$, therefore leaving us with one of 2^{2^1} cases, where the power 1 of 2^{2^1} indicates the case of one qubit:

$$\begin{aligned}
|\psi_1\rangle_0 &= \alpha|0\rangle|-\rangle_y + \beta|1\rangle|-\rangle_x \quad \text{iff } f_0(0) = 0, f_0(1) = 0,\\
|\psi_1\rangle_1 &= -i\alpha|0\rangle|+\rangle_y - \beta|1\rangle|-\rangle_x \quad \text{iff } f_1(0) = 1, f_1(1) = 1,\\
|\psi_1\rangle_2 &= \alpha|0\rangle|-\rangle_y - \beta|1\rangle|-\rangle_x \quad \text{iff } f_2(0) = 0, f_2(1) = 1,\\
|\psi_1\rangle_3 &= -i\alpha|0\rangle|+\rangle_y + \beta|1\rangle|-\rangle_x \quad \text{iff } f_3(0) = 1, f_3(1) = 0.
\end{aligned} \tag{9.1.8}$$

Once we have (9.1.8), we know simultaneously both $f(0)$ and $f(1)$ by measuring the single output state. How can we obtain (9.1.8)? Note that we cannot obtain it solely by using the usual phase kickback formation as this formation changes only the global phase and global phases are indistinguishable. For this reason, such a situation must be avoided.

Let us consider for distinguishing between the four states. Unfortunately, they are not orthogonal each other. Thus we might consider we cannot distinguish between the four states. In (9.1.8) the operations on the mapping look fine to us because the process here is based upon the phase that was obtained from the kickback formation. Therefore, the issue of orthogonality is not so essential here as we consider the phase of each state to be guaranteed.

So, by measuring $|\psi_1\rangle_i$, we can determine simultaneously all the two mappings of $f_i(x)$ for all x. Interestingly, the quantum algorithm enables us to determine a perfect property of $f_i(x)$, namely, $f_i(x)$ itself, and does it faster than a classical apparatus would. Classically namely, at least 2^2 evaluations would be necessary to that end.

Based on the above, our cryptography is as follows:

- Alice randomly selects a function f_i.
- She applies U_{f_i} to $|\psi_0\rangle$ and obtains an entangled state $|\psi_1\rangle_i$.
- She sends the entangled state $|\psi_1\rangle_i$ to Bob.
- Bob compares (by measurement) the outcome state $|\psi_1\rangle_i$ with the input state and obtains all the two mappings with the respective values for the function f_i.

- Bob learns what function Alice selected.
- Alice and Bob compare their functions (a subset of the results).
- If Eve eavesdropped, Alice and Bob would each have a different function.
- If Eve did not eavesdrop, Alice and Bob would each have the same function.

Alice and Bob perform the protocol described above many times in order to obtain enough secret keys (functions) for a secure communication.

9.2 Concrete example

We present a concrete example for a full and natural understanding of our quantum communication method. Let us consider the case where Alice selects a function f_1. Bob wants to know all the following mappings:

$$f(0) = ?, f(1) = ?. \tag{9.2.1}$$

In the classical case, Bob requires 2^2 evaluations. In the quantum case, Bob requires just one query.

Alice prepares the following input state:

$$|\psi_1\rangle_0 = \alpha|0\rangle|-\rangle_y + \beta|1\rangle|-\rangle_x \tag{9.2.2}$$

Next, Alice applies U_{f_1} to $|\psi_0\rangle$ to obtain $U_{f_1}|\psi_0\rangle = |\psi_1\rangle_1$. After that, she has the following output state:

$$|\psi_1\rangle_1 = -i\alpha|0\rangle|+\rangle_y - \beta|1\rangle|-\rangle_x \tag{9.2.3}$$

Bob enquires with Alice as to what phase factors of the quantum output state Alice has. In this example, the quantum phase factors of the output state are as follows:

$$-i, -1. \tag{9.2.4}$$

With this information, he then obtains simultaneously all the mappings of f_1:

$$f(0) = 1, f(1) = 1. \tag{9.2.5}$$

Finally, Bob learns that Alice selected a particular f_1. Again, this takes less than a classical apparatus would take, i.e., 2^2 evaluations. Likewise, Alice can select either of the 2^2 combinations of the mappings. That is, our argumentation is true for each fixed parameter i.

9.3 Quantum algorithm for determining all the three mappings of a logical function

In this section, we propose a quantum cryptography based on an algorithm for determining a function using qutrit systems. Consider a logical function f: $\{0, 1, 2\} \rightarrow \{0, 1\}$. In our protocol, Alice will know all the three mappings f(0), f(1), and f(2), that is, $f(x)$ itself. Bob will know none of them. His aim will therefore be to obtain all of them without an eavesdropper learning them. In the classical case, Bob needs 2^3 queries to learn all the mappings. In the quantum case,

Bob needs just one single query. Hence, the quantum cryptography is faster than classical cryptography by a factor of 2^3.

Quantum superposition is a fundamental feature of many quantum algorithms. It allows quantum computers to evaluate simultaneously the mappings of a function $f(x)$ for many different x. Suppose that

$$f\colon \{0, 1, 2\} \to \{0, 1\} \tag{9.3.1}$$

is a logical function known to Alice but not known to Bob. Bob's aim is therefore to determine all the mappings

$$f(0) = ?, f(1) = ?, f(2) = ?, \tag{9.3.2}$$

that is, $f(x)$ itself. In the classical case, Bob requires 2^3 queries to establish all the mappings. In the quantum case, Bob requires just a single query. Therefore, the quantum communication is faster than a classical communication which would require at least 2^3 queries.

Alice can select one of the 2^3 possible functions because of the combinations of the mappings. Later we introduce a parameter $i = 0, 1, 2, \ldots, 7$ to distinguish between these functions.

Let us discuss our quantum cryptography using qutrit systems. We introduce the transformation U_f defined by the map

$$U_f|x\rangle|j\rangle = |x\rangle|(f(x) + j) \bmod 3\rangle. \tag{9.3.3}$$

From the map U_f, we insert an imaginary number i and define the following formulas:

$$\begin{aligned} U_f|0\rangle(|0\rangle - i|1\rangle)/\sqrt{2} &= +|0\rangle(|f(0)\rangle - i|f(0) + 1\rangle)/\sqrt{2} \\ &= \begin{cases} |0\rangle(|0\rangle - i|1\rangle)/\sqrt{2} & \text{if } f(0) = 0, \\ |0\rangle(|1\rangle - i|2\rangle)/\sqrt{2} & \text{if } f(0) = 1. \end{cases} \end{aligned} \tag{9.3.4}$$

$$\begin{aligned} U_f|1\rangle(|0\rangle - |1\rangle)/\sqrt{2} &= +|1\rangle(|f(1)\rangle - |f(1) + 1\rangle)/\sqrt{2} \\ &= \begin{cases} |1\rangle(|0\rangle - |1\rangle)/\sqrt{2} & \text{if } f(1) = 0, \\ |1\rangle(|1\rangle - |2\rangle)/\sqrt{2} & \text{if } f(1) = 1. \end{cases} \end{aligned} \tag{9.3.5}$$

We define a quantum state in a three-dimensional space $|\phi\rangle$ as follows:

$$|\phi\rangle = \frac{1}{\sqrt{3}}(\omega^3|0\rangle + \omega^2|1\rangle + \omega|2\rangle), \tag{9.3.6}$$

where $\omega = e^{2\pi i/3}$. We have the following formula by the phase kickback formation:

$$U_f|2\rangle|\phi\rangle = \omega^{f(2)}|2\rangle|\phi\rangle. \tag{9.3.7}$$

In fact, from the map U_f, we can define the following formulas:

$$\begin{aligned} &U_f|2\rangle\frac{1}{\sqrt{3}}(\omega^3|0\rangle + \omega^2|1\rangle + \omega|2\rangle) \\ &= |2\rangle\frac{1}{\sqrt{3}}(\omega^3|f(2)\rangle + \omega^2|f(2)+1\rangle + \omega|f(2)+2\rangle) \\ &= \begin{cases} |2\rangle\frac{1}{\sqrt{3}}(\omega^3|0\rangle + \omega^2|1\rangle + \omega|2\rangle) & \text{if } f(2) = 0, \\ \omega|2\rangle\frac{1}{\sqrt{3}}(\omega^3|0\rangle + \omega^2|1\rangle + \omega|2\rangle) & \text{if } f(2) = 1. \end{cases} \end{aligned} \tag{9.3.8}$$

Observe that

$$(U_f)^3|x\rangle|j\rangle = |x\rangle|(3f(x)+j) \bmod 3\rangle = |x\rangle|j\rangle. \tag{9.3.9}$$

Therefore, the map U_f is a cyclic transformation. Here, we define the normalized input state ($\langle\psi_0|\psi_0\rangle = 1$) as follows:

$$\begin{aligned} &|\psi_0\rangle = \alpha|0\rangle\left[\frac{|0\rangle - i|1\rangle}{\sqrt{2}}\right] + \beta|1\rangle\left[\frac{|0\rangle - |1\rangle}{\sqrt{2}}\right] + \gamma|2\rangle|\phi\rangle, \\ &|\alpha|^2 + |\beta|^2 + |\gamma|^2 = 1, \alpha \neq 0, \beta \neq 0, \gamma \neq 0. \end{aligned} \tag{9.3.10}$$

Let us introduce a parameter i. Later, we will see that all the information for f_i is embedded into a single output state. This means that all the information for f_i can be learned from the single output state. This is the key of our quantum communication.

At the beginning of our communication protocol, Alice applies U_{f_i}, ($i = 0, 1, \ldots, 2^3 - 1$) to $|\psi_0\rangle$, $U_{f_i}|\psi_0\rangle = |\psi_1\rangle_i$, the output state is one of 2^3 cases:

$$\begin{aligned} |\psi_1\rangle_0 = \alpha|0\rangle\left[\frac{|0\rangle - i|1\rangle}{\sqrt{2}}\right] + \beta|1\rangle\left[\frac{|0\rangle - |1\rangle}{\sqrt{2}}\right] + \gamma|2\rangle|\phi\rangle \\ \text{iff } f_0(0) = 0, f_0(1) = 0, f_0(2) = 0, \end{aligned} \tag{9.3.11}$$

$$\begin{aligned} &|\psi_1\rangle_1 = \alpha|0\rangle\left[\frac{|0\rangle - i|1\rangle}{\sqrt{2}}\right] + \beta|1\rangle\left[\frac{|0\rangle - |1\rangle}{\sqrt{2}}\right] + \omega\gamma|2\rangle|\phi\rangle \\ &\text{iff } f_1(0) = 0, f_1(1) = 0, f_1(2) = 1, \end{aligned} \tag{9.3.12}$$

$$\begin{aligned} |\psi_1\rangle_2 = \alpha|0\rangle\left[\frac{|0\rangle - i|1\rangle}{\sqrt{2}}\right] + \beta|1\rangle\left[\frac{|1\rangle - |2\rangle}{\sqrt{2}}\right] + \gamma|2\rangle|\phi\rangle \\ \text{iff } f_2(0) = 0, f_2(1) = 1, f_2(2) = 0, \end{aligned} \tag{9.3.13}$$

$$|\psi_1\rangle_3 = \alpha|0\rangle\left[\frac{|0\rangle - i|1\rangle}{\sqrt{2}}\right] + \beta|1\rangle\left[\frac{|1\rangle - |2\rangle}{\sqrt{2}}\right] + \omega\gamma|2\rangle|\phi\rangle \quad \text{iff } f_3(0) = 0, f_3(1) = 1, f_3(2) = 1, \tag{9.3.14}$$

$$|\psi_1\rangle_4 = \alpha|0\rangle\left[\frac{|1\rangle - i|2\rangle}{\sqrt{2}}\right] + \beta|1\rangle\left[\frac{|0\rangle - |1\rangle}{\sqrt{2}}\right] + \gamma|2\rangle|\phi\rangle \quad \text{iff } f_4(0) = 1, f_4(1) = 0, f_4(2) = 0, \tag{9.3.15}$$

$$|\psi_1\rangle_5 = \alpha|0\rangle\left[\frac{|1\rangle - i|2\rangle}{\sqrt{2}}\right] + \beta|1\rangle\left[\frac{|0\rangle - |1\rangle}{\sqrt{2}}\right] + \omega\gamma|2\rangle|\phi\rangle \quad \text{iff } f_5(0) = 1, f_5(1) = 0, f_5(2) = 1, \tag{9.3.16}$$

$$|\psi_1\rangle_6 = \alpha|0\rangle\left[\frac{|1\rangle - i|2\rangle}{\sqrt{2}}\right] + \beta|1\rangle\left[\frac{|1\rangle - |2\rangle}{\sqrt{2}}\right] + \gamma|2\rangle|\phi\rangle \quad \text{iff } f_6(0) = 1, f_6(1) = 1, f_6(2) = 0, \tag{9.3.17}$$

$$|\psi_1\rangle_7 = \alpha|0\rangle\left[\frac{|1\rangle - i|2\rangle}{\sqrt{2}}\right] + \beta|1\rangle\left[\frac{|1\rangle - |2\rangle}{\sqrt{2}}\right] + \omega\gamma|2\rangle|\phi\rangle \quad \text{iff } f_7(0) = 1, f_7(1) = 1, f_7(2) = 1. \tag{9.3.18}$$

Let us consider for distinguishing between the eight states. Unfortunately, they are not orthogonal to each other. Thus we might consider we cannot distinguish between the eight states. In (9.3.11)–(9.3.18) the operations on the mapping look fine to us because the process here is based upon the phase that was obtained from the kickback formation. Therefore, the issue of orthogonality is not so essential here as we consider the phase of each state to be guaranteed.

By measuring the state $|\psi_1\rangle_i$ sent by Alice, Bob can determine simultaneously all the three mappings of $f_i(x)$ for all x(=0, 1, 2). Interestingly, the quantum communication gives us the ability to transmit a perfect property of $f_i(x)$, namely, $f_i(x)$ itself. Moreover, the quantum transmission is faster than classical communication which would require 2^3 queries.

With the above, our cryptography is as follows:

- Alice selects a function f_i at random.
- She applies U_{f_i} to $|\psi_0\rangle$ and obtains an entangled state $|\psi_1\rangle_i$.
- She sends the entangled state $|\psi_1\rangle_i$ to Bob.
- Bob compares (by measurement) the result state $|\psi_1\rangle_i$ with the input state and obtains all the three mappings with regards to the function f_i.
- Bob learns what function Alice selected.
- Alice and Bob compare their functions (a subset of the results).
- If Eve eavesdropped, Alice and Bob would each have a different function.
- If Eve did not eavesdrop, Alice and Bob would each have the same function.

Alice and Bob perform the protocol described above many times in order to obtain enough secret keys (functions).

9.4 Concrete example

For a full and natural understanding of our quantum communication method, we present below a concrete example for qutrit systems. Let us consider the case where Alice selects a function f_1. Bob wants to know all the mappings

$$f(0) = ?, f(1) = ?, f(2) = ?. \tag{9.4.1}$$

In the classical case, Bob requires 2^3 evaluations. In the quantum case, Bob requires just one query.

At the beginning, Alice prepares the following input state:

$$|\psi_0\rangle = \alpha|0\rangle\left[\frac{|0\rangle - i|1\rangle}{\sqrt{2}}\right] + \beta|1\rangle\left[\frac{|0\rangle - |1\rangle}{\sqrt{2}}\right] + \gamma|2\rangle|\phi\rangle. \tag{9.4.2}$$

Next, Alice applies U_{f_1} to $|\psi_0\rangle$ obtaining $U_{f_1}|\psi_0\rangle = |\psi_1\rangle_1$. Her output state is

$$|\psi_1\rangle_1 = \alpha|0\rangle\left[\frac{|0\rangle - i|1\rangle}{\sqrt{2}}\right] + \beta|1\rangle\left[\frac{|0\rangle - |1\rangle}{\sqrt{2}}\right] + \omega\gamma|2\rangle|\phi\rangle. \tag{9.4.3}$$

Bob enquires with Alice as to what phase factors of the quantum output state Alice has. In this example, the quantum phase factors of the output state are as follows:

$$1, 1, \omega. \tag{9.4.4}$$

Then Bob obtains simultaneously all the mappings of f_1:

$$f(0) = 0, f(1) = 0, f(2) = 1. \tag{9.4.5}$$

Finally, Bob learns that Alice selected the mapping f_1. Again, the quantum method is faster than a classical apparatus which would require 2^3 evaluations. Likewise, Alice can select any of the 2^3 combinations of the mappings. That is, our argumentation holds for each fixed parameter i.

9.5 Quantum algorithm for determining all the 2^2 mappings of a logical function

In this section, we propose a quantum cryptography based on an algorithm for determining a function using qubit systems. Consider the logical function f: $\{0, 1\}^2 \to \{0, 1\}$. Assume that Alice knows all the 2^2 mappings $f(0, 0)$, $f(0, 1)$, $f(1, 0)$, and $f(1, 1)$, that is, $f(x)$ itself. Assume further that Bob knows none of them. His aim is then to obtain all of these mapping values without an eavesdropper learning them. In the classical case, Bob needs 2^{2^2} queries. In the quantum case, Bob needs just a single query. Thus, the quantum cryptography is faster than classical cryptography by a factor of 2^{2^2}.

We propose a quantum algorithm for determining the 2^2 mappings of a function. Suppose that

$$f: \{0, 1\}^2 \to \{0, 1\} \tag{9.5.1}$$

is a logical function. We want to know simultaneously the 2^2 mappings $f(0, 0)$, $f(0, 1)$, $f(1, 0)$, and $f(1, 1)$. Later we will see a complete match between our results and a Boolean algebra F_2. In the Boolean algebra F_2, the functions are of two variables. For example, $f(x, y)$ is the function where x and y are variables used in mapping f. In what follows, the abbreviation $f(xy)$ will stand for $f(x, y)$.

We define the input state as follows using an application of (9.1.7):

$$\begin{aligned} &|\psi_0\rangle = a_1|00\rangle|-\rangle_y + a_2|01\rangle|-\rangle_y + a_3|10\rangle|-\rangle_x + a_4|11\rangle|-\rangle_x, \\ &\langle\psi_0|\psi_0\rangle = 1 \Leftarrow |a_1|^2 + |a_2|^2 + |a_3|^2 + |a_4|^2 = 1, \\ &a_1 \neq 0,\ a_2 \neq 0,\ a_3 \neq 0,\ a_4 \neq 0. \end{aligned} \tag{9.5.2}$$

From the mapping U_f, we can define the following formulas:

$$U_f|00\rangle|-\rangle_y = \begin{cases} (-i)^{f(00)}|00\rangle|-\rangle_y & \text{if } f(00) = 0, \\ (-i)^{f(00)}|00\rangle|+\rangle_y & \text{if } f(00) = 1. \end{cases} \tag{9.5.3}$$

$$U_f|01\rangle|-\rangle_y = \begin{cases} (-i)^{f(01)}|01\rangle|-\rangle_y & \text{if } f(01) = 0, \\ (-i)^{f(01)}|01\rangle|+\rangle_y & \text{if } f(01) = 1. \end{cases} \tag{9.5.4}$$

$$U_f|10\rangle|-\rangle_x = \begin{cases} (-1)^{f(10)}|10\rangle|-\rangle_x & \text{if } f(10) = 0, \\ (-1)^{f(10)}|10\rangle|-\rangle_x & \text{if } f(10) = 1. \end{cases} \tag{9.5.5}$$

$$U_f|11\rangle|-\rangle_x = \begin{cases} (-1)^{f(11)}|11\rangle|-\rangle_x & \text{if } f(11) = 0, \\ (-1)^{f(11)}|11\rangle|-\rangle_x & \text{if } f(11) = 1. \end{cases} \tag{9.5.6}$$

Applying U_{f_i}, $(i = 0, 1, 2, \ldots, 2^{2^2} - 1)$, to $|\psi_0\rangle$ gives $U_{f_i}|\psi_0\rangle = |\psi_1\rangle_i$ and leaves us with one of the 2^{2^2} cases:

$$\begin{aligned} &|\psi_1\rangle_0 = a_1|00\rangle|-\rangle_y + a_2|01\rangle|-\rangle_y + a_3|10\rangle|-\rangle_x + a_4|11\rangle|-\rangle_x \\ &\text{iff } f_0(00) = 0, f_0(01) = 0, f_0(10) = 0, f_0(11) = 0, \end{aligned} \tag{9.5.7}$$

$$\begin{aligned} &|\psi_1\rangle_1 = a_1|00\rangle|-\rangle_y + a_2|01\rangle|-\rangle_y + a_3|10\rangle|-\rangle_x - a_4|11\rangle|-\rangle_x \\ &\text{iff } f_1(00) = 0, f_1(01) = 0, f_1(10) = 0, f_1(11) = 1, \end{aligned} \tag{9.5.8}$$

$$\begin{aligned} &|\psi_1\rangle_2 = a_1|00\rangle|-\rangle_y + a_2|01\rangle|-\rangle_y - a_3|10\rangle|-\rangle_x + a_4|11\rangle|-\rangle_x \\ &\text{iff } f_2(00) = 0, f_2(01) = 0, f_2(10) = 1, f_2(11) = 0, \end{aligned} \tag{9.5.9}$$

$$|\psi_1\rangle_3 = a_1|00\rangle|-\rangle_y + a_2|01\rangle|-\rangle_y - a_3|10\rangle|-\rangle_x - a_4|11\rangle|-\rangle_x$$
$$\text{iff } f_3(00) = 0, f_3(01) = 0, f_3(10) = 1, f_3(11) = 1, \tag{9.5.10}$$

$$|\psi_1\rangle_4 = a_1|00\rangle|-\rangle_y - ia_2|01\rangle|+\rangle_y + a_3|10\rangle|-\rangle_x + a_4|11\rangle|-\rangle_x$$
$$\text{iff } f_4(00) = 0, f_4(01) = 1, f_4(10) = 0, f_4(11) = 0, \tag{9.5.11}$$

$$|\psi_1\rangle_5 = a_1|00\rangle|-\rangle_y - ia_2|01\rangle|+\rangle_y + a_3|10\rangle|-\rangle_x - a_4|11\rangle|-\rangle_x$$
$$\text{iff } f_5(00) = 0, f_5(01) = 1, f_5(10) = 0, f_5(11) = 1, \tag{9.5.12}$$

$$|\psi_1\rangle_6 = a_1|00\rangle|-\rangle_y - ia_2|01\rangle|+\rangle_y - a_3|10\rangle|-\rangle_x + a_4|11\rangle|-\rangle_x$$
$$\text{iff } f_6(00) = 0, f_6(01) = 1, f_6(10) = 1, f_6(11) = 0, \tag{9.5.13}$$

$$|\psi_1\rangle_7 = a_1|00\rangle|-\rangle_y - ia_2|01\rangle|+\rangle_y - a_3|10\rangle|-\rangle_x - a_4|11\rangle|-\rangle_x$$
$$\text{iff } f_7(00) = 0, f_7(01) = 1, f_7(10) = 1, f_7(11) = 1, \tag{9.5.14}$$

and so on. Finally, we have

$$|\psi_1\rangle_{15} = -ia_1|00\rangle|+\rangle_y - ia_2|01\rangle|+\rangle_y - a_3|10\rangle|-\rangle_x - a_4|11\rangle|-\rangle_x$$
$$\text{iff } f_{15}(00) = 1, f_{15}(01) = 1, f_{15}(10) = 1, f_{15}(11) = 1, \tag{9.5.15}$$

Let us consider for distinguishing between the 16 states. Unfortunately, they are not orthogonal each other. Thus we might consider we cannot distinguish between the 16 states. In (9.5.7)–(9.5.15) the operations on the mapping look fine to us because the process here is based upon the phase obtained from the kickback formation. So, the issue of orthogonality is not so essential because we consider the phase of each state to be guaranteed here.

By measuring $|\psi_1\rangle_i$ we can determine simultaneously all the 2^2 mappings of $f_i(x, y)$ for all x and y. Interestingly, the quantum algorithm gives us the ability to determine a perfect property of $f_i(x, y)$, namely, $f_i(x, y)$ itself. This determination is faster than with a classical apparatus which would require at least 2^{2^2} evaluations.

Our cryptography is as follows:

- Alice randomly selects a function f_i.
- She applies U_{f_i} to $|\psi_0\rangle$ and obtains an entangled state $|\psi_1\rangle_i$.
- She sends the entangled state $|\psi_1\rangle_i$ to Bob.
- Bob compares (by measurement) the result state $|\psi_1\rangle_i$ with the input state and obtains all the four mappings with the values concerning the function f_i.
- Bob learns what function Alice selected.
- Alice and Bob compare their functions (a subset of the results).
- If Eve eavesdropped, Alice and Bob would each have a different function.
- If Eve did not eavesdrop, Alice and Bob would each have the same function.

Alice and Bob perform the protocol described above many times in order to obtain enough secret keys (functions).

9.6 Concrete example

Let us consider the case where Alice selects a function f_1. Bob wants to know all the following mappings:

$$f(0, 0) = ?, f(0, 1) = ?, f(1, 0) = ?, f(1, 1) = ?. \tag{9.6.1}$$

In the classical case, Bob requires 2^{2^2} evaluations. In the quantum case, Bob requires just one query.

Alice prepares the following input state:

$$|\psi_0\rangle = a_1|00\rangle|-\rangle_y + a_2|01\rangle|-\rangle_y + a_3|10\rangle|-\rangle_x + a_4|11\rangle|-\rangle_x. \tag{9.6.2}$$

Next, Alice applies U_{f_1} to $|\psi_0\rangle$ to obtain $U_{f_1}|\psi_0\rangle = |\psi_1\rangle_1$. She has the following output state:

$$|\psi_1\rangle_1 = a_1|00\rangle|-\rangle_y + a_2|01\rangle|-\rangle_y + a_3|10\rangle|-\rangle_x - a_4|11\rangle|-\rangle_x. \tag{9.6.3}$$

Bob enquires with Alice as to what phase factors of the quantum output state Alice has. In this example, the quantum phase factors of the output state are as follows:

$$1, 1, 1, -1. \tag{9.6.4}$$

Then, Bob obtains simultaneously all the mappings of f_1:

$$f(0, 0) = 0, f(0, 1) = 0, f(1, 0) = 0, f(1, 1) = 1. \tag{9.6.5}$$

Finally, Bob realizes that Alice selected f_1. Again, this quantum communication is faster than using a classical apparatus which would require 2^{2^2} evaluations. Likewise, Alice can select any of the 2^{2^2} combinations of the mappings. That is, our argumentation holds for each fixed parameter i.

In conclusion, we have studied a quantum cryptography based on an algorithm for determining all the mappings of a logical function simultaneously using an entangled state. The security of our cryptography is based on the Ekert 1991 protocol, which uses an entangled state. Consequently, eavesdropping destroyed the entanglement. In the cryptography, Alice selected a secret function among the possible function types. Bob's aim was then to determine the selected function (a key) without an eavesdropper learning it. In order for both Alice and Bob to be able to select the same function classically, in the worst case Bob would require multiple queries to Alice. In the quantum case, however, Bob required just a single query. By measuring the single entangled state, which was sent to him by Alice, Bob obtained the function that Alice had selected. This quantum key distribution method is faster than the multiple classical queries that would be required in the classical case.

Parts of this chapter have been reprinted from Nagata K, Wong R, Ngoc Diep D and Nakamura T 2021 Quantum cryptography based on an algorithm for determining simultaneously all the mappings of a Boolean function *Cryptology* ePrint Archive 758 (iacr.org)

Koji Nagata, Do Ngoc Diep, Ahmed Farouk and Tadao Nakamura

Chapter 10

Quantum cryptography based on an algorithm for determining a function using qudit systems

In this chapter, we propose quantum cryptography based on an algorithm for determining a function using qudit entangled states. The security of our cryptography is based on the Ekert 1991 protocol, that is, we use an entangled state. An eavesdropper must destroy the entangled state. Consider one of many functions. Alice selects a secret function. Bob's aim is of determining the selected function (a key) without the eavesdropper knowing it. To select the same function, in the classical case, in the worst case, Bob needs many queries to Alice. In the quantum case, Bob needs just one query. By measuring the single entangled state, which is sent by Alice, Bob can obtain the selected function. This is faster than classical cryptography.

10.1 Quantum cryptography based on an algorithm for determining a function using qudit systems

Quantum superposition is a fundamental feature of many quantum algorithms. It allows quantum computers to evaluate simultaneously the mappings of a function $f(x)$ for many different x. Suppose

$$f\colon \{0, 1, 2, \dots, d-1\} \to \{0, 1, 2, \dots, e-1\}, (d \geqslant e) \tag{10.1.1}$$

is a function. We use the condition (10.1.1) in order to make the function $\omega^{f(x)}$ to be univalent, where $\omega = e^{2\pi i/d}$. There are e^d functions. Alice selects a function secretly. Bob's aim is of determining the secret function, that is, $f(x)$ itself without Eve's attack. To select the same function, in the classical case, in the worst case, Bob requires e^d queries to Alice. In the quantum case, Bob requires just one query. This is faster than classical cryptography by a factor of e^d.

Alice can select one of the e^d functions because of the combinations of the mappings. Later we introduce a parameter $i = 0, 1, 2, \dots, e^d - 1$ for the functions.

doi:10.1088/978-0-7503-4700-6ch10

Let us discuss our quantum cryptography using qudit systems. We introduce the transformation U_f defined by the mapping

$$U_f|x\rangle|j\rangle = |x\rangle|(f(x) + j) \bmod d\rangle. \tag{10.1.2}$$

We define a quantum state in a d-dimensional space $|\phi_d\rangle$ as follows:

$$|\phi_d\rangle = \frac{1}{\sqrt{d}}(\omega^d|0\rangle + \omega^{d-1}|1\rangle + \cdots + \omega|d-1\rangle), \tag{10.1.3}$$

where $\omega = e^{2\pi i/d}$. We have the following formula by the phase kickback formation:

$$U_f|x\rangle|\phi_d\rangle = \omega^{f(x)}|x\rangle|\phi_d\rangle. \tag{10.1.4}$$

Notice

$$(U_f)^d|x\rangle|j\rangle = |x\rangle|(df(x) + j) \bmod d\rangle = |x\rangle|j\rangle. \tag{10.1.5}$$

Therefore, the mapping U_f is a cyclic transformation. Here, we define the normalized input state ($\langle\psi_0|\psi_0\rangle = 1$) as follows:

$$|\psi_0\rangle = \sum_{n=1}^{d} \alpha_n|n-1\rangle|\phi_n\rangle, \quad \sum_{n=1}^{d}|\alpha_n|^2 = 1, \alpha_1 \neq 0, \alpha_2 \neq 0, \ldots, \alpha_d \neq 0. \tag{10.1.6}$$

Let us introduce a parameter i. Later, we see all the information for f_i is embedded into a single output entangled state. This means Bob gets all the information for f_i when he knows the single output entangled state. This is the key of our quantum cryptography.

Alice applies U_{f_i}, $(i = 0, 1, \ldots, e^d - 1)$ to $|\psi_0\rangle$, $U_{f_i}|\psi_0\rangle = |\psi_1\rangle_i$, the output entangled state is one of e^d cases:

$$|\psi_1\rangle_i = \sum_{n=1}^{d} \omega^{f_i(n-1)}\alpha_n|n-1\rangle|\phi_n\rangle, \text{ iff } f_i(n-1) = 0, 1, 2, \ldots, \text{ or } e-1. \tag{10.1.7}$$

So, by measuring an entangled state $|\psi_1\rangle_i$, which is sent by Alice, Bob may determine the secret function. This is very interesting indeed: our quantum cryptography gives us the ability to transmit a perfect property of $f_i(x)$, namely, $f_i(x)$ itself without Eve's attack. This is faster than classical cryptography by a factor of e^d.

Our cryptography is as follows:

- Alice randomly selects a function f_i.
- She applies U_{f_i} to $|\psi_0\rangle$ in giving an entangled state $|\psi_1\rangle_i$.
- She sends the entangled state $|\psi_1\rangle_i$ to Bob.
- Bob compares (by measurement) the result state $|\psi_1\rangle_i$ with the input state and obtains all the d mappings with the values concerning the function f_i.
- Bob realizes what function Alice selects.
- Alice and Bob compare their functions (subset of the results).
- If Eve's attack exists, Alice and Bob select the different function.
- If Eve's attack does not exist, Alice and Bob select the same function.

Alice and Bob perform the protocol described above many times of obtaining enough secret keys (functions).

10.2 Concrete example

We present a concrete example to understand our quantum cryptography fully and naturally. Let us consider the case where Alice randomly selects a function f_1. Bob wants to know the secret function without Eve's attack. In the classical case, in the worst case, Bob requires e^d evaluations. In the quantum case, Bob requires just a query.

Alice prepares the following input entangled state:

$$|\psi_0\rangle = \sum_{n=1}^{d} \alpha_n |n - 1\rangle |\phi_n\rangle. \tag{10.2.1}$$

Next, Alice applies U_{f_1} to $|\psi_0\rangle$, $U_{f_1}|\psi_0\rangle = |\psi_1\rangle_1$. She has, for example, the following output entangled state:

$$\begin{aligned} |\psi_1\rangle_1 &= \sum_{n=1}^{d} \omega^{f_1(n-1)} \alpha_n |n - 1\rangle |\phi_n\rangle \\ &\text{iff } f_1(d - 1) = 1, f_1(n) = 0, n = 0, 1, 2, \dots, d - 2. \end{aligned} \tag{10.2.2}$$

Bob asks what quantum output entangled state Alice has. Then Bob obtains simultaneously all the mappings with the values of $f_1(x)$:

$$f_1(0) = 0, f_1(1) = 0, f_1(2) = 0, \dots, f_1(d - 1) = 1. \tag{10.2.3}$$

Bob realizes that Alice selects $f_1(x)$. Alice and Bob compare their functions (subset of the results). If Eve's attack exists, Alice and Bob select the different function. If Eve's attack does not exist, Alice and Bob select the same function. Alice and Bob perform the protocol described above of obtaining enough secret keys (functions) many times.

Again, this is faster than classical cryptography which would require in the worst case e^d evaluations. Likewise, Alice can select the e^d combinations of the mappings. That is, our argumentations are true for each parameter i.

Parts of this chapter have been reprinted from Nagata K, Diep D N and Nakamura T 2020 Quantum cryptography based on an algorithm for determining a function using qudit systems *Int. J. Theor. Phys.* **59** 2875.

IOP Publishing

Simplified Quantum Computing with Applications

Koji Nagata, Do Ngoc Diep, Ahmed Farouk and Tadao Nakamura

Chapter 11

Continuous-variable quantum computing and its applications to cryptography

In this chapter, we propose quantum cryptography based on an algorithm for determining a function using continuous-variable entangled states. As the security of our cryptography is based on Ekert's 1991 protocol, we use an entangled state. An eavesdropper will destroy the entangled state. Consider the very large number of possible functions. Alice selects a secret function. Bob's aim is to determine the selected function (a key) without the eavesdropper learning it. Classically, in order to select the same function, Bob would require a very large number of queries to Alice in the worst-case scenario. In the quantum case, Bob requires just a single query. By measuring the single entangled state, which is sent to him by Alice, Bob can obtain the selected function. This protocol's performance is faster than the very large number of queries required in the case of classical cryptography.

11.1 Quantum cryptography based on an algorithm for determining a function using continuous-variable entangled states

Suppose f: $[0, d] \rightarrow [0, e]$, $(1 \leqslant e \leqslant d)$, $e, d \in \mathbf{N}$ is a function. We specify the condition $(e \leqslant d)$ in order to make the function $\omega^{f(x)}$ univalent, where $\omega = e^{2\pi i/d}$. There is a very large number of functions of this form. Alice selects a function $f(x)$ secretly. Bob's aim is then to determine the secret function $f(x)$ without Eve's interference. Classically, in the worst case, Bob needs to query Alice a very large number of times in order to be able to select the same function as Alice has. Example queries would be, e.g., 'What is the value of $f(0.2)$?', 'What is the value of $f(1)$?', and so on. In the quantum case, however, Bob needs just a single query, which is faster than the classical scenario of having to query Alice a very large number of times.

Alice can select one of the very large number of functions. Later we will introduce a continuous parameter $i \in [0, e^d]$ for the functions f_i.

doi:10.1088/978-0-7503-4700-6ch11

Let us discuss our quantum cryptography using continuous-variable entangled states. To that end, we introduce the transformation U_f defined by the mapping $U_f|x\rangle|j\rangle = |x\rangle|(f(x) + j) \bmod d\rangle$. We define a quantum state $|\phi_d\rangle$ in an infinite-dimensional space as follows: $|\phi_d\rangle = \int_{j\in[0,+d)} dj \frac{\omega^{d-j}|j\rangle}{\sqrt{d}}$, where $\omega = e^{2\pi i/d}$. By the phase kickback formation, we have the following formula: $U_f|x\rangle|\phi_d\rangle = \omega^{f(x)}|x\rangle|\phi_d\rangle$. Notice that $(U_f)^d|x\rangle|j\rangle = |x\rangle|(df(x) + j) \bmod d\rangle = |x\rangle|j\rangle$. Therefore, the mapping of U_f is a cyclic transformation. Here, we define the normalized input state $(\langle\psi_0|\psi_0\rangle = 1)$ as follows: $|\psi_0\rangle = \int_0^d dn\alpha(n)|n\rangle|\phi_n\rangle$, $\int_0^d dn|\alpha(n)|^2 = 1$, $\alpha(n) \neq 0$.

Now, let us introduce the continuous parameter i. Later, we will see that all the information for f_i is embedded in a single output entangled state. Therefore, knowing the single output entangled state, Bob will be able to obtain all the information for f_i. This is the key of our quantum cryptography.

By applying U_{f_i}, $(i \in [0, e^d])$, to $|\psi_0\rangle$ Alice obtains the following output entangled state: $U_{f_i}|\psi_0\rangle = |\psi_1\rangle_i = \int_0^d dn\omega^{f_i(n)}\alpha(n)|n\rangle|\phi_n\rangle$ iff $f_i(n) \in [0, e]$, $\forall n \in [0, d]$.

So, by measuring the entangled state $|\psi_1\rangle_i$, which he received from Alice, Bob will be able to determine the secret function that Alice selected. Interestingly, our quantum cryptography gives us the ability to transmit a perfect property of $f_i(x)$, namely, the $f_i(x)$ itself, without Eve's interference. This is faster than endlessly querying Alice, which would have been the case classically.

Our cryptography is as follows:

- Alice selects a function f_i, $i \in [0, e^d]$, at random.
- Alice applies U_{f_i} to $|\psi_0\rangle$, which results in an entangled state $|\psi_1\rangle_i$.
- Alice sends the entangled state $|\psi_1\rangle_i$ to Bob.
- Bob compares (by measurement) the state $|\psi_1\rangle_i$ with the input state and obtains all the mappings for the values of the function f_i.
- Bob determines what function Alice has selected.
- Alice and Bob compare their respective functions (subset of the results).
- If Eve has interfered, Alice and Bob determine that they each have a different function.
- If Eve hasn't interfered, Alice and Bob determine that they share the same function.

Alice and Bob carry out the protocol described above many times to share enough secret keys (functions). Again, this protocol is faster than the very large querying executed in the corresponding classical cryptography.

In what follows, we consider a concrete example.

11.2 Concrete example

We present a concrete example to facilitate a full and natural understanding of our quantum cryptography. Let us consider the case where Alice randomly selects a secret function $f_0(x) = x$ and we assume that $d = e$. Bob wants to learn the secret function without Eve's interference. Classically, in the worst case, Bob would need to query Alice a very large number of times. In the quantum case, however, Bob requires just a single query.

Alice prepares the following input entangled state: $|\psi_0\rangle = \int_0^d dn\alpha(n)|n\rangle|\phi_n\rangle$. Then, she applies U_{f_0} to $|\psi_0\rangle$ and obtains the following output entangled state: $U_{f_0}|\psi_0\rangle = |\psi_1\rangle_0 = \int_0^d dn\omega^{f_0(n)}\alpha(n)|n\rangle|\phi_n\rangle$ iff $f_0(n) = n \in [0, d]$. Upon Bob's inquiry as to what quantum output state she has obtained, Alice sends the entangled state $|\psi_1\rangle_0$ to Bob over a quantum channel. Then Bob obtains simultaneously all the mappings with the values $f_0(x) = x$, $\forall x \in [0, d]$. Based on that, Bob determines that Alice has selected $f_0(x) = x$.

Alice and Bob execute the protocol described above a number of times to obtain enough secret keys (functions) changing the secret function. After that, Alice and Bob compare their functions (subset of the results). If Eve's interference is established, Alice and Bob will determine that they have different functions. If Eve hasn't interfered, Alice and Bob will determine that they share the same function.

Again, this protocol is faster than the corresponding classical cryptography, which would require, in the worst case, a very large number of queries. Likewise, Alice can select among the very large number of combinations of mappings. Our argument is true for every parameter i.

Parts of this chapter have been reprinted from Diep D N, Nagata K and Wong R 2020 Continuous-variable quantum computing and its applications to cryptography *Int. J. Theor. Phys.* **59** 3184.

Koji Nagata, Do Ngoc Diep, Ahmed Farouk and Tadao Nakamura

Chapter 12

Various new forms of the Bernstein–Vazirani algorithm beyond qubit systems

In this chapter, we present various new forms of the generalized Bernstein–Vazirani algorithm beyond qubit systems. We hope all the generalizations presented here would improve quantum algorithm science.

12.1 Algorithm for determining a bit string

First, we review the generalized Bernstein–Vazirani algorithm for determining a bit string. The algorithm has the feature of the Hadamard transform of a quantum state in a two-dimensional space to solve the problem of finding a bit string.

We suppose that the following sequence of complex numbers is given

$$a_1, a_2, a_3, \dots, a_N. \tag{12.1.1}$$

We introduce a special function

$$g: \mathbf{C} \to \{0, 1\}. \tag{12.1.2}$$

Our goal is for determining the following values (as a bit string)

$$g(a_1), g(a_2), g(a_3), \dots, g(a_N). \tag{12.1.3}$$

Recall that in the best classical case, we need 2^N queries, that is, 2^N separate evaluations of the function (12.1.2). In the case of the Bernstein–Vazirani algorithm, we shall require a single query.

We define

$$g(a) = (g(a_1), g(a_2), g(a_3), \dots, g(a_N)), \tag{12.1.4}$$

where each entry of $g(a)$ is a bit. Here $g(a) \in \{0, 1\}^N$. We define $f(x)$ as follows:

$$\begin{aligned} f(x) &= (g(a) \cdot x) \bmod 2 \\ &= \{g(a_1)x_1 + g(a_2)x_2 + \cdots + g(a_N)x_N\} \bmod 2, \end{aligned} \tag{12.1.5}$$

doi:10.1088/978-0-7503-4700-6ch12

where $x = (x_1, \dots, x_N)$ and $x_j \in \{0, 1\}$. The entries of x and $g(a_j)$ are in $\{0, 1\}$. Let us follow the quantum states through our algorithm.

The input state is

$$|\psi_0\rangle = |0\rangle^{\otimes N}|1\rangle \tag{12.1.6}$$

where $|0\rangle^{\otimes N} = \overbrace{|0\rangle \otimes |0\rangle \otimes \cdots \otimes |0\rangle}^{N}$. After the componentwise Hadamard transforms on the state (12.1.6)

$$\overbrace{H|0\rangle \otimes H|0\rangle \otimes \cdots \otimes H|0\rangle}^{N} \otimes H|1\rangle \tag{12.1.7}$$

we have, for quantum gates and their theory,

$$|\psi_1\rangle = \sum_{x\in\{0,1\}^N} \frac{|x\rangle}{\sqrt{2^N}}\left[\frac{|0\rangle - |1\rangle}{\sqrt{2}}\right]. \tag{12.1.8}$$

Next, the function f is evaluated using

$$U_f|x, y\rangle = |x, y \oplus f(x)\rangle \tag{12.1.9}$$

in giving

$$|\psi_2\rangle = (-1)^{f(x)}|\psi_1\rangle. \tag{12.1.10}$$

Here $y \oplus f(x)$ is the bitwise XOR (exclusive OR) of y and $f(x)$. By checking the cases $x_1 = 0$ and $x_1 = 1$ separately, we see that for a single qubit

$$H|x_1\rangle = \sum_{z_1}(-1)^{x_1 z_1}|z_1\rangle/\sqrt{2}. \tag{12.1.11}$$

Thus we have

$$H^{\otimes N}|x_1, \dots, x_N\rangle = \frac{\sum_{z_1,\dots,z_N} (-1)^{\{x_1z_1+\cdots+x_Nz_N\}(mod2)}|z_1, \dots, z_N\rangle}{\sqrt{2^N}}. \tag{12.1.12}$$

Here, $x_j z_j$ is in modulo 2. This can be summarized more succinctly in the very useful equation

$$H^{\otimes N}|x\rangle = \frac{\sum_z (-1)^{x\cdot z}|z\rangle}{\sqrt{2^N}} \tag{12.1.13}$$

where $x \cdot z$ is the bitwise inner product of x and z, modulo 2. Using the equation (12.1.10) and (12.1.13), we can now evaluate $H^{\otimes N}|\psi_2\rangle = |\psi_3\rangle$

$$|\psi_3\rangle = \sum_z \sum_x \frac{(-1)^{x\cdot z + f(x)}|z\rangle}{2^N}\left[\frac{|0\rangle - |1\rangle}{\sqrt{2}}\right]. \tag{12.1.14}$$

Thus we have

$$|\psi_3\rangle = \sum_z \sum_x \frac{(-1)^{x\cdot z + g(a)\cdot x}|z\rangle}{2^N}\left[\frac{|0\rangle - |1\rangle}{\sqrt{2}}\right]. \tag{12.1.15}$$

We can see that

$$\sum_x (-1)^{x\cdot z + g(a)\cdot x} = 2^N \delta_{g(a),z}. \tag{12.1.16}$$

Therefore, the sum is zero if $z \neq g(a)$ and is 2^N if $z = g(a)$. Thus we have

$$\begin{aligned} |\psi_3\rangle &= \sum_z \sum_x \frac{(-1)^{x\cdot z + g(a)\cdot x}|z\rangle}{2^N}\left[\frac{|0\rangle - |1\rangle}{\sqrt{2}}\right] \\ &= \sum_z \frac{2^N \delta_{g(a),z}|z\rangle}{2^N}\left[\frac{|0\rangle - |1\rangle}{\sqrt{2}}\right] \\ &= |(g(a_1), g(a_2), \cdots, g(a_N))\rangle\left[\frac{|0\rangle - |1\rangle}{\sqrt{2}}\right] \end{aligned} \tag{12.1.17}$$

from which

$$|(g(a_1), g(a_2), \cdots, g(a_N))\rangle \tag{12.1.18}$$

can be obtained. That is to say, if we measure $|(g(a_1), g(a_2), \cdots, g(a_N))\rangle$ then we can retrieve the following values (as a bit string)

$$g(a_1), g(a_2), g(a_3), \ldots, g(a_N) \tag{12.1.19}$$

using a single query. The step of determining the string of bits is shown to outperform its classical case by a factor of 2^N.

12.2 Extension to a natural number string

Second, we discuss the generalized Bernstein–Vazirani algorithm for determining a natural number string. The generalized algorithm presented here has the feature of the normal Fourier transform of a quantum state in a finite-dimensional space to solve the problem of finding a string of natural numbers. The result is the most notable generalization.

Let us discuss the Bernstein–Vazirani algorithm of determining a natural number string.

We suppose that the following sequence of complex numbers is given

$$a_1, a_2, a_3, \ldots, a_N. \tag{12.2.1}$$

We introduce a special function

$$g: \mathbf{C} \to \{0, 1, 2, 3, \ldots, d-1\}. \tag{12.2.2}$$

Our aim is to determine the values below as a string of natural numbers,

$$g(a_1),\ g(a_2),\ g(a_3),\ \dots,\ g(a_N). \tag{12.2.3}$$

Recall that in the best classical case, we need d^N queries, that is, d^N separate evaluations of the function (12.2.2). In the case of the generalized Bernstein–Vazirani algorithm, we shall require a single query.

We introduce a positive integer d. Throughout the discussion, we consider the problem in modulo d. Assume the following

$$0 \leqslant \overbrace{g(a_1),\ g(a_2),\ g(a_3),\ \dots,\ g(a_N)}^{N} \leqslant d-1, \tag{12.2.4}$$

where $g(a_j) \in \{0, 1, \dots, d-1\}$, and we define

$$g(a) = (g(a_1),\ g(a_2),\ g(a_3),\ \dots,\ g(a_N)), \tag{12.2.5}$$

where each entry of $g(a)$ is a natural number. Here $g(a) \in \{0, 1, \dots, d-1\}^N$. We define $f(x)$ as follows:

$$\begin{aligned} f(x) &= (g(a)\cdot x) \bmod d \\ &= \{g(a_1)x_1 + g(a_2)x_2 + \cdots + g(a_N)x_N\} \bmod d, \end{aligned} \tag{12.2.6}$$

where $x = (x_1, \dots, x_N) \in \{0, 1, \dots, d-1\}^N$. Let us follow the quantum states through the algorithm.

The input state is

$$|\psi_0\rangle = |0\rangle^{\otimes N}|d-1\rangle, \tag{12.2.7}$$

where $|0\rangle^{\otimes N}$ means $|\overbrace{0,0,\dots,0}^{N}\rangle$. Here $|0\rangle$ and $|d-1\rangle$ are quantum states in a d-dimensional space. We discuss the Fourier transform of $|0\rangle$

$$F|0\rangle = \sum_{y=0}^{d-1} \frac{|y\rangle}{\sqrt{d}}. \tag{12.2.8}$$

We define a quantum state in a d-dimensional space $|\phi\rangle$ as follows:

$$|\phi\rangle = \frac{1}{\sqrt{d}}(\omega^d|0\rangle + \omega^{d-1}|1\rangle + \cdots + \omega|d-1\rangle), \tag{12.2.9}$$

where $\omega = e^{2\pi i/d}$. In the following, we discuss the Fourier transform of $|d-1\rangle$

$$F|d-1\rangle = |\phi\rangle. \tag{12.2.10}$$

The Fourier transform of $|x_1 \dots x_N\rangle$ is as follows:

$$F^{\otimes N}|x_1 \dots x_N\rangle = \sum_{z\in K} \frac{\omega^{z\cdot x}|z\rangle}{\sqrt{d^N}}, \tag{12.2.11}$$

where $K = \{0, 1, \dots, d-1\}^N$ and z is $(z_1, z_2, \dots, z_N)$. Here, for completeness, $\sum_{z\in K}$ is a shorthand to the compound sum

$$\sum_{z_1\in\{0,1,\dots,d-1\}} \cdots \sum_{z_N\in\{0,1,\dots,d-1\}}. \tag{12.2.12}$$

After the componentwise Fourier transforms of the first N quantum states in a d-dimensional space and after the Fourier transform of the quantum state in a d-dimensional space $|d-1\rangle$ in (12.2.7)

$$\overbrace{F|0\rangle \otimes F|0\rangle \otimes \cdots \otimes F|0\rangle}^{N} \otimes F|d-1\rangle, \tag{12.2.13}$$

we have, for quantum gates and their theory,

$$|\psi_1\rangle = \sum_{x \in K} \frac{|x\rangle}{\sqrt{d^N}} |\phi\rangle. \tag{12.2.14}$$

Here, the notation $F|0\rangle$ means the Fourier transform of $|0\rangle$ and the notation $F|d-1\rangle$ means the Fourier transform of $|d-1\rangle$.

We introduce $U_{f(x)}$ gate

$$U_{f(x)}|x\rangle|j\rangle = |x\rangle|(f(x)+j) \bmod d\rangle, \tag{12.2.15}$$

where

$$f(x) = g(a) \cdot x \bmod d. \tag{12.2.16}$$

We have the following formula by the phase kickback formation:

$$U_{f(x)}|x\rangle|\phi\rangle = \omega^{f(x)}|x\rangle|\phi\rangle. \tag{12.2.17}$$

We have $|\psi_2\rangle$, by operating $U_{f(x)}$ to $|\psi_1\rangle$,

$$U_{f(x)}|\psi_1\rangle = |\psi_2\rangle = \sum_{x \in K} \frac{\omega^{f(x)}|x\rangle}{\sqrt{d^N}} |\phi\rangle. \tag{12.2.18}$$

After the Fourier transform of $|x\rangle$, using the previous equations (12.2.11) and (12.2.18), we can now evaluate $|\psi_3\rangle$ as follows:

$$|\psi_3\rangle = \sum_{z \in K} \sum_{x \in K} \frac{(\omega)^{x \cdot z + g(a) \cdot x}|z\rangle}{d^N} |\phi\rangle. \tag{12.2.19}$$

Notice

$$\sum_{x \in K} (\omega)^{x \cdot (z + g(a))} = d^N \delta_{z, \vec{d} - g(a)}, \tag{12.2.20}$$

where $\vec{d} = \overbrace{(d, d, \ldots, d)}^{N}$. Therefore, the above summation is zero if $z \neq \vec{d} - g(a)$ and the above summation is d^N if $z = \vec{d} - g(a)$. Thus we have

$$\begin{aligned} |\psi_3\rangle &= \sum_{z \in K} \frac{d^N \delta_{z, \vec{d} - g(a)}|z\rangle}{d^N} |\phi\rangle \\ &= |\vec{d} - (g(a_1), g(a_2), g(a_3), \ldots, g(a_N))\rangle|\phi\rangle \end{aligned} \tag{12.2.21}$$

from which

$$|\vec{d} - (g(a_1), g(a_2), g(a_3), \dots, g(a_N))\rangle \tag{12.2.22}$$

can be obtained. That is to say, if we measure the first N quantum states in a d-dimensional space of the state $|\psi_3\rangle$, that is, $|\vec{d} - (g(a_1), g(a_2), g(a_3), \dots, g(a_N))\rangle$, then we can retrieve the following values (as a natural number string)

$$g(a_1), g(a_2), g(a_3), \dots, g(a_N) \tag{12.2.23}$$

using a single query. The step of determining the string of natural numbers is shown to outperform its classical case by a factor of d^N.

12.3 Extension to an integer string

Thirdly, we discuss the generalized Bernstein–Vazirani algorithm for determining an integer string. The generalized algorithm presented here has the feature of a general discrete non-unitary transform of a quantum state in a finite-dimensional space to solve the problem of finding a string of integers.

We present an algorithm for determining the values of a function that are extended to integers. That is, the extended algorithm determines an integer string instead of a natural number string. This is a different point. Here, we use the general non-unitary transform instead of the Fourier transform. The transform is a straightforward generalization of the Fourier transform.

We suppose that the following sequence of complex numbers is given

$$a_1, a_2, a_3, \dots, a_N. \tag{12.3.1}$$

We introduce a special function

$$g: \mathbf{C} \to \{-(d-1), \dots, -2, -1, 0, 1, 2, \dots, d-1\}. \tag{12.3.2}$$

Our aim is to determine the values below as a string of natural numbers,

$$g(a_1), g(a_2), g(a_3), \dots, g(a_N). \tag{12.3.3}$$

Recall that in classical case, we need $(2d-1)^N$ queries, that is, $(2d-1)^N$ separate evaluations of the function (12.3.2). In the case of the generalized Bernstein–Vazirani algorithm, we shall require a single query.

We introduce a positive integer d. Throughout the discussion, we consider the problem in modulo d. Assume the following

$$-(d-1) \leqslant \overbrace{g(a_1), g(a_2), g(a_3), \dots, g(a_N)}^{N} \leqslant d-1, \tag{12.3.4}$$

where $g(a_j) \in \{-(d-1), \dots, -1, 0, 1, \dots, d-1\}$, and we define

$$g(a) = (g(a_1), g(a_2), g(a_3), \dots, g(a_N)), \tag{12.3.5}$$

where each entry of $g(a)$ is an integer. Here $g(a) \in \{-(d-1), \dots, -1, 0, 1, \dots, d-1\}^N$. We define $f(x)$ as follows:

$$\begin{aligned} f(x) &= (g(a) \cdot x) \bmod d \\ &= \{g(a_1)x_1 + g(a_2)x_2 + \cdots + g(a_N)x_N\} \bmod d, \end{aligned} \tag{12.3.6}$$

where $x = (x_1, \ldots, x_N) \in \{-(d-1), \ldots, -1, 0, 1, \ldots, d-1\}^N$. Let us follow the quantum states through the algorithm.

The input state is

$$|\psi_0\rangle = |0\rangle^{\otimes N}|d-1\rangle, \tag{12.3.7}$$

where $|0\rangle^{\otimes N}$ means $\overbrace{|0,0,\ \ldots,\ 0\rangle}^{N}$. Here $|0\rangle$ is a quantum state in a $(2d-1)$-dimensional space and $|d-1\rangle$ is a quantum state in a d-dimensional space. We discuss the general transform of $|0\rangle$

$$G|0\rangle = \sum_{y=-(d-1)}^{d-1} \frac{|y\rangle}{\sqrt{2d-1}}. \tag{12.3.8}$$

A calculation shows that this is not a unitary operation; (in fact, it does not have full rank).

We define a quantum state in a d-dimensional space $|\phi\rangle$ as follows:

$$|\phi\rangle = \frac{1}{\sqrt{d}}(\omega^d|0\rangle + \omega^{d-1}|1\rangle + \cdots + \omega|d-1\rangle), \tag{12.3.9}$$

where $\omega = e^{2\pi i/d}$. In the following, we discuss the Fourier transform of $|d-1\rangle$

$$F|d-1\rangle = |\phi\rangle. \tag{12.3.10}$$

The general transform of $|x_1 \ldots x_N\rangle$ is as follows:

$$G^{\otimes N}|x_1 \ldots x_N\rangle = \sum_{z \in K} \frac{\omega^{z \cdot x}|z\rangle}{\sqrt{(2d-1)^N}}, \tag{12.3.11}$$

where $K = \{-(d-1), \ldots, -1, 0, 1, \ldots, d-1\}^N$ and z is $(z_1, z_2, \ldots, z_N)$. $z_j \in \{-(d-1), \ldots, -1, 0, 1, \ldots, d-1\}$. This is not a unitary operation. Here, for completeness, $\sum_{z \in K}$ is a shorthand to the compound sum

$$\sum_{z_1 \in \{-(d-1), \ldots, -1, 0, 1, \ldots, d-1\}} \cdots \sum_{z_N \in \{-(d-1), \ldots, -1, 0, 1, \ldots, d-1\}}. \tag{12.3.12}$$

After the componentwise general transforms of the first N quantum states in a $(2d-1)$-dimensional space and after the Fourier transform of the quantum state in a d-dimensional space $|d-1\rangle$ in (12.3.7)

$$\overbrace{G|0\rangle \otimes G|0\rangle \otimes \cdots \otimes G|0\rangle}^{N} \otimes F|d-1\rangle, \tag{12.3.13}$$

we have, for quantum gates and their theory,

$$|\psi_1\rangle = \sum_{x \in K} \frac{|x\rangle}{\sqrt{(2d-1)^N}} |\phi\rangle. \tag{12.3.14}$$

Here, the notation $G|0\rangle$ means the general transform of $|0\rangle$ and the notation $F|d-1\rangle$ means the Fourier transform of $|d-1\rangle$.

We introduce $U_{f(x)}$ gate

$$U_{f(x)}|x\rangle|j\rangle = |x\rangle|(f(x)+j) \bmod d\rangle, \tag{12.3.15}$$

where

$$f(x) = g(a) \cdot x \bmod d. \tag{12.3.16}$$

We have the following formula by the phase kickback formation:

$$U_{f(x)}|x\rangle|\phi\rangle = \omega^{f(x)}|x\rangle|\phi\rangle. \tag{12.3.17}$$

We have $|\psi_2\rangle$, by operating $U_{f(x)}$ to $|\psi_1\rangle$,

$$U_{f(x)}|\psi_1\rangle = |\psi_2\rangle = \sum_{x \in K} \frac{\omega^{f(x)}|x\rangle}{\sqrt{(2d-1)^N}} |\phi\rangle. \tag{12.3.18}$$

After the general transform of $|x\rangle$, using the previous equations (12.2.10) and (12.3.18), we can now evaluate $|\psi_3\rangle$ as follows:

$$|\psi_3\rangle = \sum_{z \in K} \sum_{x \in K} \frac{(\omega)^{x \cdot z + g(a) \cdot x}|z\rangle}{(2d-1)^N} |\phi\rangle. \tag{12.3.19}$$

Notice

$$\sum_{x \in K} (\omega)^{x \cdot (z+g(a))} = (2d-1)^N \delta_{z,-g(a)}. \tag{12.3.20}$$

Therefore, the above summation is zero if $z \neq -g(a)$ and the above summation is $(2d-1)^N$ if $z = -g(a)$. Thus we have

$$\begin{aligned} |\psi_3\rangle &= \sum_{z \in K} \frac{(2d-1)^N \delta_{z,-g(a)}|z\rangle}{(2d-1)^N} |\phi\rangle \\ &= |-(g(a_1), g(a_2), g(a_3), \ldots, g(a_N))\rangle|\phi\rangle \end{aligned} \tag{12.3.21}$$

from which

$$|-(g(a_1), g(a_2), g(a_3), \ldots, g(a_N))\rangle \tag{12.3.22}$$

can be obtained. That is to say, if we measure the first N quantum states in a $(2d-1)$-dimensional space of the state $|\psi_3\rangle$, that is, $|-(g(a_1), g(a_2), g(a_3), \ldots, g(a_N))\rangle$, then we can retrieve the following values (as an integer string)

$$g(a_1), g(a_2), g(a_3), \ldots, g(a_N) \tag{12.3.23}$$

using a single query. The step of determining the string of integers is shown to outperform its classical case by a factor of $(2d - 1)^N$.

12.4 Extension to a complex number string

Finally, we discuss the generalized Bernstein–Vazirani algorithm for determining a complex number string. The generalized algorithm presented here has the feature of a general continuous non-unitary transform of a quantum state in an infinite-dimensional space to solve the problem of finding a string of complex numbers.

Suppose that the following sequence of complex numbers is given

$$a_1, a_2, a_3, \ldots, a_N. \tag{12.4.1}$$

A special function is provided as follows:

$$g: \mathbf{C} \to \mathbf{C}. \tag{12.4.2}$$

Our final goal is for determining the following values as a complex number string

$$g(a_1), g(a_2), g(a_3), \ldots, g(a_N). \tag{12.4.3}$$

It is worth pointing out that in the best classical case, to determine $g(a_1), \ldots, g(a_N)$, a large number of queries are expected to be answered, which are a large number of separate evaluations of the function (12.4.2). However, in the case of the generalized Bernstein–Vazirani algorithm, we shall require a single query instead of a large number of queries.

We determine the real part. Assume the following sequence of complex numbers as follows:

$$a_1, a_2, a_3, \ldots, a_N. \tag{12.4.4}$$

Take a special function l as follows:

$$l: \mathbf{C} \to (-d, +d). \tag{12.4.5}$$

Our aim is for determining the following values as the real part of the complex number string

$$l(a_1), l(a_2), l(a_3), \ldots, l(a_N). \tag{12.4.6}$$

Recall that in the best classical case, to determine $l(a_1), \ldots, l(a_N)$ as the coefficients of the linear function $f(x) = l(a_1)x_1 + \cdots + l(a_N)x_N$, we need a large number queries for values of $f(x)$, that is, a large number of separate evaluations of the function (12.4.5). Using the generalized Bernstein–Vazirani algorithm, we require just a single query.

Considering a positive integer as d, the problem is discussed in modulo d:

$$-d < \overbrace{l(a_1), l(a_2), l(a_3), \ldots, l(a_N)}^{N} < d, \tag{12.4.7}$$

where $l(a_j) \in (-d, +d)$, and $l(a)$ is as follows:

$$l(a) = (l(a_1), l(a_2), l(a_3), \ldots, l(a_N)), \tag{12.4.8}$$

where each entry of $l(a)$ is a real number, and $l(a) \in (-d, +d)^N$. We define $f(x)$ as follows:

$$\begin{aligned} f(x) &= (l(a) \cdot x) \bmod d \\ &= \{l(a_1)x_1 + l(a_2)x_2 + \cdots + l(a_N)x_N\} \bmod d, \end{aligned} \tag{12.4.9}$$

where $x = (x_1, \ldots, x_N) \in (-d, +d)^N$. Let us follow the quantum states through the algorithm.

The input state is

$$|\psi_0\rangle = |0\rangle^{\otimes N}|\phi\rangle, \tag{12.4.10}$$

where $|0\rangle^{\otimes N}$ means $\overbrace{|0,0, \ldots, 0\rangle}^{N}$. Here $N < \infty$ that is N is a finite natural number. $|\phi\rangle$ is defined as follows:

$$|\phi\rangle = \sum_{j\in[0,+d)} \frac{\omega^{d-j}|j\rangle}{\sqrt{d}} \tag{12.4.11}$$

where $\omega = e^{2\pi i/d}$.

The general transform of $|0\rangle$ is as follows:

$$G|0\rangle = \sum_{y\in(-d,+d)} \frac{|y\rangle}{\sqrt{2d}}. \tag{12.4.12}$$

A calculation shows that this is not a unitary operation. Indeed, it does not have a full rank.

The general transform of $|x_1 \ldots x_N\rangle$ is as follows:

$$G^{\otimes N}|x_1 \ldots x_N\rangle = \sum_{z\in K} \frac{\omega^{z\cdot x}|z\rangle}{\sqrt{(2d)^N}}, \tag{12.4.13}$$

where $K = (-d, d)^N$ and z is $(z_1, z_2, \ldots, z_N)$. $z_j \in (-d, +d)$. This is not a unitary operation. Here, for completeness, $\sum_{z\in K}$ is written in shorthand to a compound sum as:

$$\sum_{z_1\in(-d,+d)} \cdots \sum_{z_N\in(-d,+d)}. \tag{12.4.14}$$

After the componentwise general transforms of the first N quantum states in an infinite-dimensional space in (12.4.10)

$$\overbrace{G|0\rangle \otimes G|0\rangle \otimes \cdots \otimes G|0\rangle}^{N} \otimes |\phi\rangle, \tag{12.4.15}$$

we have, for quantum gates and their theory,

$$|\psi_1\rangle = \sum_{x\in K} \frac{|x\rangle}{\sqrt{(2d)^N}}|\phi\rangle. \tag{12.4.16}$$

Here, the notation $G|0\rangle$ means the general transform of $|0\rangle$. It is a straightforward generalization of the Fourier transform.

We introduce $U_{f(x)}$ gate

$$U_{f(x)}|x\rangle|j\rangle = |x\rangle|(f(x) + j) \bmod d\rangle, \tag{12.4.17}$$

where

$$f(x) = (l(a) \cdot x) \bmod d. \tag{12.4.18}$$

We have the following formula by the phase kickback formation:

$$U_{f(x)}|x\rangle|\phi\rangle = \omega^{f(x)}|x\rangle|\phi\rangle. \tag{12.4.19}$$

In what follows, we discuss the rationale of the above relation (12.4.19). Now consider the action of the $U_{f(x)}$ gate to the state $|x\rangle|\phi\rangle$. Each term in $|\phi\rangle$ is of the form $\omega^{d-j}|j\rangle$. We see

$$U_{f(x)}\omega^{d-j}|x\rangle|j\rangle = \omega^{d-j}|x\rangle|(j + f(x)) \bmod d\rangle. \tag{12.4.20}$$

k is introduced such as $f(x) + j = k$ then $d - j = d + f(x) - k$. Consequently, (12.4.20) becomes

$$U_{f(x)}\omega^{d-j}|x\rangle|j\rangle = \omega^{f(x)}\omega^{d-k}|x\rangle|k \bmod d\rangle. \tag{12.4.21}$$

Now, when $k < d$, we have $|k \bmod d\rangle = |k\rangle$ and thus, the terms in $|\phi\rangle$ such that $k < d$ are transformed as follows:

$$U_{f(x)}\omega^{d-j}|x\rangle|j\rangle = \omega^{f(x)}\omega^{d-k}|x\rangle|k\rangle. \tag{12.4.22}$$

Furthermore, as $f(x)$ and j are both bounded above by d, k is strictly less than $2d$. Thus, when $d \leqslant k < 2d$, we have $|k \bmod d\rangle = |k - d\rangle$. Now, introducing m as $k - d = m$, we have

$$\omega^{f(x)}\omega^{d-k}|x\rangle|k \bmod d\rangle = \omega^{f(x)}\omega^{-m}|x\rangle|m\rangle = \omega^{f(x)}\omega^{d-m}|x\rangle|m\rangle. \tag{12.4.23}$$

Hence, the terms in $|\phi\rangle$ such that $k \geqslant d$ are transformed as follows:

$$U_{f(x)}\omega^{d-j}|x\rangle|j\rangle = \omega^{f(x)}\omega^{d-m}|x\rangle|m\rangle. \tag{12.4.24}$$

Subsequently, regarding (12.4.22) and (12.4.24), we have

$$U_{f(x)}|x\rangle|\phi\rangle = \omega^{f(x)}|x\rangle|\phi\rangle. \tag{12.4.25}$$

Therefore, the relation (12.4.19) holds. Operating $U_{f(x)}$ to $|\psi_1\rangle$, $|\psi_2\rangle$ is transformed as:

$$U_{f(x)}|\psi_1\rangle = |\psi_2\rangle = \sum_{x\in K} \frac{\omega^{f(x)}|x\rangle}{\sqrt{(2d)^N}}|\phi\rangle. \tag{12.4.26}$$

After the general transform of $|x\rangle$, using the previous equations (12.2.10) and (12.4.26), $|\psi_3\rangle$ can be evaluated as follows:

$$|\psi_3\rangle = \sum_{z\in K}\sum_{x\in K}\frac{(\omega)^{x\cdot z + l(a)\cdot x}|z\rangle}{(2d)^N}|\phi\rangle. \tag{12.4.27}$$

Notice

$$\sum_{x\in K}(\omega)^{x\cdot(z+l(a))} = (2d)^N\delta_{z,-l(a)}. \tag{12.4.28}$$

Therefore, the above sum is zero if $z \neq -l(a)$ and the above sum is $(2d)^N$ if $z = -l(a)$. So, we have

$$\begin{aligned}|\psi_3\rangle &= \sum_{z\in K}\frac{(2d)^N\delta_{z,-l(a)}|z\rangle}{(2d)^N}|\phi\rangle \\ &= |-(l(a_1), l(a_2), l(a_3), \dots, l(a_N))\rangle|\phi\rangle\end{aligned} \tag{12.4.29}$$

from which

$$|-(l(a_1), l(a_2), l(a_3), \dots, l(a_N))\rangle \tag{12.4.30}$$

can be obtained. That is to say, if we measure the first N quantum states in an infinite-dimensional space of the state $|\psi_3\rangle$, that is, $|-(l(a_1), l(a_2), l(a_3), \dots, l(a_N))\rangle$, then we can retrieve the following values as the real part of a complex number strings

$$l(a_1), l(a_2), l(a_3), \dots, l(a_N) \tag{12.4.31}$$

using a single query.

Next, we determine the imaginary part. We suppose that the following sequence of complex numbers is given

$$a_1, a_2, a_3, \dots, a_N. \tag{12.4.32}$$

We introduce a special function

$$h: \mathbf{C} \rightarrow (-d, +d). \tag{12.4.33}$$

Our aim is to determine the following values as the imaginary part of a complex number string,

$$h(a_1), h(a_2), h(a_3), \dots, h(a_N). \tag{12.4.34}$$

Recall that in the best classical case, to determine $h(a_1), \dots, h(a_N)$, we need a large number of queries, that is, a large number of separate evaluations of the function (12.4.33). In the case of the generalized Bernstein–Vazirani algorithm, we shall require a single query.

Finally, the two evaluations (the real part and the imaginary part) can be performed, in a parallel computing method, simultaneously. In the final step, we measure the following quantum state:

$$|-(l(a_1), l(a_2), l(a_3), \dots, l(a_N))\rangle \otimes |-(h(a_1), h(a_2), h(a_3), \dots, h(a_N))\rangle. \tag{12.4.35}$$

That is, we determine the N complex values $g(a_j) = l(a_j) + ih(a_j)$ simultaneously. The step of determining the string of complex numbers is shown to outperform its classical case by a large number factor.

12.5 Extension to a matrix string

Additionally, we propose a method for calculating simultaneously many different matrices A, B, C, ... into $g(A)$, $g(B)$, $g(C)$, ... The step of solving the problem is shown to outperform the classical case by a large number factor.

We propose a method for calculating simultaneously many different matrices A, B, C, ... into $g(A)$, $g(B)$, $g(C)$, ... That is, the generalized Bernstein–Vazirani algorithm determines a matrix string instead of a number string. The step of solving the problem is shown to outperform the classical case by a large number factor.

Let the jth matrix be a $N_j \times M_j$ matrix (N_j rows, M_j columns, and $j = 1, 2, 3, \ldots$). To simplify, let us consider the case of two matrices ($j = 2$). Given the elements of the matrices $a_1, a_2, a_3, \ldots, a_{N_1 \times M_1}$,

$b_1, b_2, b_3, \ldots, b_{N_2 \times M_2}$, and a special function g newly, we calculate simultaneously $O(N_1 \times M_1 + N_2 \times M_2)$ values of the function $g(a_1), g(a_2), g(a_3), \ldots, g(a_{N_1 \times M_1})$,

$g(b_1), g(b_2), g(b_3), \ldots, g(b_{N_2 \times M_2})$. In the classical case, we have to query a large number factor. However, in the quantum case, we need just a query. Thus, the step of calculating the two matrices is shown to outperform the classical case by a large number factor.

The generalization of it, to many matrices, can be done in a straightforward manner. The most significant result of this section is that it can be better than its classical counterpart regarding computational complexity. The step of calculating the different matrices is shown to outperform the classical case by a large number factor.

Let us consider the case of a single $N \times M$ matrix A (N rows and M columns). Given the elements of the matrix $a_1, a_2, a_3, \ldots, a_{N \times M}$, and a special function g newly, we can obtain the following values (as a number string)

$$g(a_1), g(a_2), g(a_3), \ldots, g(a_{N \times M}) \tag{12.5.1}$$

using a single query. This is the most significant point of our method and it is possible by the generalized Bernstein–Vazirani algorithm. We consider them as the elements of the matrix $g(A)$;

$$g(A) = \begin{pmatrix} a_{11} & a_{12} & \ldots & a_{1m} \\ a_{21} & a_{22} & \ldots & a_{2m} \\ \vdots & \vdots & & \vdots \\ a_{n1} & a_{n2} & \ldots & a_{nm} \end{pmatrix}, \tag{12.5.2}$$

where $(g(a_1), g(a_2), g(a_3), \ldots, g(a_{N \times M})) = (a_{11}, a_{21}, \ldots, a_{nm})$. Therefore, we obtain $g(A)$ by a single query. In the classical case, we obtain $g(A)$ by a large number of queries. Thus, the step of calculating the matrix $g(A)$ is shown to outperform the classical case by a large number factor.

Let us consider the case of two matrices A and B (N_j rows, M_j columns, and $j = 1, 2$). Given the elements of the matrices $a_1, a_2, a_3, \ldots, a_{N_1 \times M_1}, b_1, b_2, b_3, \ldots, b_{N_2 \times M_2}$, and a special function g newly, we can obtain the following values (as a number string)

$$g(a_1), g(a_2), g(a_3), \ldots, g(a_{N_1 \times M_1}), g(b_1), g(b_2), g(b_3), \ldots, g(b_{N_2 \times M_2}), \tag{12.5.3}$$

using a single query. Again it is possible by the generalized Bernstein–Vazirani algorithm. We consider them as the elements of the matrices $g(A)$ and $g(B)$;

$$g(A) = \begin{pmatrix} a_{11} & a_{12} & \ldots & a_{1m_1} \\ a_{21} & a_{22} & \ldots & a_{2m_1} \\ \vdots & \vdots & & \vdots \\ a_{n_1 1} & a_{n_1 2} & \ldots & a_{n_1 m_1} \end{pmatrix}, \tag{12.5.4}$$

and

$$g(B) = \begin{pmatrix} b_{11} & b_{12} & \ldots & b_{1m_2} \\ b_{21} & b_{22} & \ldots & b_{2m_2} \\ \vdots & \vdots & & \vdots \\ b_{n_2 1} & b_{n_2 2} & \ldots & b_{n_2 m_2} \end{pmatrix}, \tag{12.5.5}$$

where $(g(a_1), g(a_2), g(a_3), \ldots, g(a_{N_1 \times M_1})) = (a_{11}, a_{21}, \ldots, a_{n_1 m_1})$. $(g(b_1), g(b_2), g(b_3), \ldots, g(b_{N_2 \times M_2})) = (b_{11}, b_{21}, \ldots, b_{n_2 m_2})$. Therefore, we obtain $g(A)$ and $g(B)$ by a single query. In the classical case, we obtain $g(A)$ and $g(B)$ by a large number of queries. Thus, the step of calculating the matrix $g(A)$ and $g(B)$ is shown to outperform the classical case by a large number factor.

The generalization of it, to many matrices, can be done in a straightforward manner. The step of calculating the matrices is shown to outperform its classical case by a large number factor.

In what follows, we give an example in the case of three matrices. Let $g(x)$ be x^2. Let A, B, C be

$$A = \begin{pmatrix} 0 & 2 \\ 1 & 3 \end{pmatrix}, B = \begin{pmatrix} 4 & 6 \\ 5 & 7 \end{pmatrix}, C = \begin{pmatrix} 8 & 10 \\ 9 & 11 \end{pmatrix}. \tag{12.5.6}$$

The elements are

$$\overbrace{0,1,\ 2,3}^{A},\ \overbrace{4,5,\ 6,7}^{B},\ \overbrace{8,9,\ 10,11}^{C}. \tag{12.5.7}$$

Recall that in the classical case, we need d^N queries, that is, d^N separate evaluations of the function $g(x) = x^2$ ($d = 12$ and $N = 12$).

$$g(0), g(1), g(2), g(3), g(4), g(5), g(6), g(7), g(8), g(9), g(10), g(11). \tag{12.5.8}$$

In the quantum case, from the generalized Bernstein–Vazirani algorithm, using g, we have

$$\overbrace{0{,}1,\ 4{,}9}^{g(A)},\ \overbrace{16{,}25,\ 36{,}49}^{g(B)},\ \overbrace{64{,}81,\ 100{,}121}^{g(C)} \tag{12.5.9}$$

using a single query. Hence we have

$$g(A) = \begin{pmatrix} 0 & 4 \\ 1 & 9 \end{pmatrix},\ g(B) = \begin{pmatrix} 16 & 36 \\ 25 & 49 \end{pmatrix},\ g(C) = \begin{pmatrix} 64 & 100 \\ 81 & 121 \end{pmatrix}, \tag{12.5.10}$$

using a single query. The step of calculating the matrices is shown to outperform the classical case by a factor of d^N ($d = 12$ and $N = 12$).

Now, we generalize the Bernstein–Vazirani algorithm from a bit string into a matrix string. Let us suppose that the following sequence of matrices is given

$$A_1,\ A_2,\ A_3,\ \ldots\ ,\ A_N. \tag{12.5.11}$$

Let us now introduce a special function newly

$$g\colon \mathbf{C} \to \mathbf{C}. \tag{12.5.12}$$

Our goal is for determining the following matrices (as a matrix string)

$$g(A_1),\ g(A_2),\ g(A_3),\ \ldots\ ,\ g(A_N). \tag{12.5.13}$$

In the generalized Bernstein–Vazirani algorithm, we shall require a single query. We hope our discussions will give a way to the quantum simulation problem.

Integers are more general than natural numbers. Complex numbers are more general than integers. In fact, we solve many problems by using complex numbers and we cannot answer many problems without them. Hence, we definitely need complex numbers. Thus, we want to generalize the Bernstein–Vazirani algorithm for determining the values of a special function when the values are complex numbers because the utility of it becomes much wider than previous ones. For example, we would find simultaneously all the roots in the root finding problem. Also, we can plot a graph of a given function immediately.

Quantum mechanics is formulated by the matrix theory. So it is useful to treat matrices by quantum computing if we want to simulate quantum phenomena by such a computer. For example, it is desirable if quantum computers calculate highly complicated quantum chemistry phenomena. Now, we generalize the Bernstein–Vazirani algorithm from a bit string into a matrix string. We hope our discussions will give a first step to the quantum simulation problem.

Parts of this chapter have been reprinted from Nagata K, Patro S K, Geurdes H, Heidari S, Diep D N and Nakamura T 2019 Various new forms of the Bernstein-Vazirani algorithm beyond qubit systems *Asian J. Math. Phys.* **3** 1 (https://mathphys.asia/index.php/volume-3-2019/)

Koji Nagata, Do Ngoc Diep, Ahmed Farouk and Tadao Nakamura

Chapter 13

Creating genuine quantum algorithms for quantum energy-based computing

In this chapter, we consider a particular interpretation of quantum mechanics in which we assume that the quantum under consideration is energy. We discuss an algorithm based on this energy interpretation. We present a method for rapidly determining a homogeneous linear function $f(x) := s \cdot x = s_1 x_1 + s_2 x_2 + \cdots + s_N x_N$, where $x = (x_1, \dots, x_N)$, $x_j \in \mathbf{R}$ and the coefficients $s = (s_1, \dots, s_N)$, $s_j \in \mathbf{Z}$. Given the interpolation values $(f(1), f(2), \dots, f(N)) = \vec{y}$, we shall determine simultaneously the unknown coefficients $s = (s_1(\vec{y}), \dots, s_N(\vec{y}))$ of the linear function. The step of determining the values is shown to outperform the classical equivalent by a factor of $(2d - 1)^N$. Our method is based on the generalized Bernstein–Vazirani algorithm for qudit systems. Next, by using M parallel quantum systems, we find simultaneously M homogeneous linear functions. The step of obtaining the set of M homogeneous linear functions is shown to outperform the classical case by a factor of $(2d - 1)^N \times M$.

13.1 Quantum algorithm for determining a homogeneous linear function

Suppose that f is a homogeneous linear function $f(x) := s \cdot x = s_1 x_1 + s_2 x_2 + \cdots + s_N x_N$. Here $x = (x_1, \dots, x_N)$, $x_j \in \mathbf{R}$ and the coefficients $s = (s_1, \dots, s_N)$, $s_j \in \mathbf{Z}$. Our goal is of determining the unknown coefficients $s_1,\dots, s_N$ from simultaneous knowledge of the interpolation values $f(1), f(2), \dots, f(N)$. Because the function is linear, we need to know exactly N points $(1, f(1)),\dots,(N, f(N))$ to interpolate the function in the classical case, i.e., we need $(2d - 1)^N$ computational steps. However, as we show, in the quantum mechanical case, we need only one query.

Here, we are given the interpolation values $f(1), f(2), \dots, f(N)$

doi:10.1088/978-0-7503-4700-6ch13

$$\begin{aligned} f(1) &= s_1 a_1 + s_2 a_2 + \cdots + s_N a_N = y^1, \\ f(2) &= s_1 b_1 + s_2 b_2 + \cdots + s_N b_N = y^2, \\ &\vdots \\ f(N) &= s_1 c_1 + s_2 c_2 + \cdots + s_N c_N = y^N. \end{aligned} \tag{13.1.1}$$

Given the values $\vec{y} = (y^1, y^2, \ldots y^N)$, our aim is of determining simultaneously the following values:

$$s = (s_1(\vec{y}), s_2(\vec{y}), \ldots, s_N(\vec{y})) = (s_1, \ldots, s_N). \tag{13.1.2}$$

According to Kronecker's theorem, the system (13.1.1) of linear equations has a unique solution $s = (s_1, \ldots, s_N)$ if and only if the augmented coefficient matrix

$$(A|\mathbf{b}) := \begin{pmatrix} a_1 & a_2 & \ldots & a_N & y^1 \\ b_1 & b_2 & \ldots & b_N & y^2 \\ \vdots & \vdots & & \vdots & \vdots \\ c_1 & c_2 & \ldots & c_N & y^N \end{pmatrix} \tag{13.1.3}$$

has rank N, i.e., the interpolation points $(a, y^1), (b, y^2), \ldots, (c, y^N)$ are in generic position. This problem can be solved using the algorithm for determining an integer string in the modulo d system. Here we assume

$$\max\{s_1, \ldots, s_N\} \leqslant d - 1. \tag{13.1.4}$$

Hence, we arrive at our goal of finding the unknown coefficients $s = (s_1, \ldots, s_N)$ from the given $(f(1), f(2), \ldots, f(N))$. Thus, we know the unknown coefficients $s = (s_1, \ldots, s_N)$ of the linear function $f(x) = s \cdot x$, if we know the interpolation values $(f(1), f(2), \ldots, f(N))$.

13.2 Quantum algorithm for determining M homogeneous linear functions

By using M parallel quantum systems, we have simultaneously M homogeneous linear functions. Suppose that $f^1, f^2, \ldots, f^M$ are M homogeneous linear functions

$$\begin{aligned} f^1(x) &= s_1^1 x_1 + s_2^1 x_2 + \cdots + s_N^1 x_N, \\ f^2(x) &= s_1^2 x_1 + s_2^2 x_2 + \cdots + s_N^2 x_N, \\ &\vdots \\ f^M(x) &= s_1^M x_1 + s_2^M x_2 + \cdots + s_N^M x_N. \end{aligned} \tag{13.2.1}$$

Here $x = (x_1, \ldots, x_N)$, $x_j \in \mathbf{R}$, and the coefficients $s_j^k \in \mathbf{Z}$. We have to be given the set of interpolation values

$$\begin{aligned} \vec{y}^1 &= (y_1^1, y_1^2, \ldots, y_1^N), \\ \vec{y}^2 &= (y_2^1, y_2^2, \ldots, y_2^N), \\ &\vdots \\ \vec{y}^M &= (y_M^1, y_M^2, \ldots, y_M^N). \end{aligned} \tag{13.2.2}$$

Our aim is of determining simultaneously the coefficients

$$\begin{aligned}
&s_1^1(\vec{y}^1), s_2^1(\vec{y}^1), \ldots, s_N^1(\vec{y}^1) = s_1^1, \ldots, s_N^1, \\
&s_1^2(\vec{y}^2), s_2^2(\vec{y}^2), \ldots, s_N^2(\vec{y}^2) = s_1^2, \ldots, s_N^2, \\
&\vdots \\
&s_1^M(\vec{y}^M), s_2^M(\vec{y}^M), \ldots, s_N^M(\vec{y}^M) = s_1^M, \ldots, s_N^M.
\end{aligned} \tag{13.2.3}$$

This problem can be solved by the algorithm using many parallel quantum systems. In this case, we measure the following quantum state

$$\begin{aligned}
&|-(s_1^1(\vec{y}^1), s_2^1(\vec{y}^1), \ldots, s_N^1(\vec{y}^1))\rangle \otimes | - (s_1^2(\vec{y}^2), s_2^2(\vec{y}^2), \ldots, s_N^2(\vec{y}^2))\rangle\otimes \\
&\cdots \otimes |-(s_1^M(\vec{y}^M), s_2^M(\vec{y}^M), \ldots, s_N^M(\vec{y}^M))\rangle.
\end{aligned} \tag{13.2.4}$$

That is to say, if we measure the quantum state (13.2.4), then we can retrieve the coefficients of the M functions with a single query. The step of obtaining the set of M homogeneous linear functions is shown to outperform the classical case by a factor of $(2d - 1)^N \times M$.

The double-slit experiment is an illustration of wave–particle duality—a concept that many find hard to understand. Let us consider the double-slit experiment from the point of view of energy. In this experiment, a source emits a beam of energy in the form of photons, which are particles. The beam travels through a barrier in which there are two slits. If one places a detector screen on the other side of the barrier, the pattern of detected energy exhibits interference fringes, which are characteristic of waves, although the detector actually responds to the energy of particles. Thus, the system exhibits behavior that although it can be interpreted both in terms of waves (interference patterns) and in terms of particles (dots on the screen), is basically related to energy. In this energy interpretation, a quantum state is a distribution of energy. Thus, quantum measurement theory involves investigating distributions of energy. This approach allows us to create genuine quantum algorithms.

Quantum computers as currently conceived are actually 'quantum estimators,' and there are many problems that cannot be solved using such estimators even though they are solvable using classical approaches. Therefore, there is a need for more rigorous quantum devices to deal with such problems. With this as our motivation, we shall discuss an arbitrarily high-dimensional quantum computer that should be capable of solving a wide range of problems as a first step toward the development of a true quantum computer.

Parts of this chapter have been reprinted from Nagata K, Nakamura T, Geurdes H, Batle J, Abdalla S, Farouk A and Diep D N 2018 Creating very true quantum algorithms for quantum energy based computing *Int. J. Theor. Phys.* **57** 973

Koji Nagata, Do Ngoc Diep, Ahmed Farouk and Tadao Nakamura

Chapter 14

Quantum algorithms for finding the roots of a polynomial function

In this chapter, we consider quantum algorithms for finding the roots of a polynomial function.

14.1 Finding the roots of a polynomial function by using a bit string

In this section, an approach in order to find the roots of a polynomial function is investigated. The method is based on the Bernstein–Vazirani algorithm generalized to the modulo 2 for determining a bit string.

First, the polynomial function $f(x)$ can be defined as follows:

$$f(x) = x^m + a_{m-1}x^{m-1} + \cdots + a_1x + a_0. \tag{14.1.1}$$

Here all the roots are in the integers $\mathbf{Z}$.

Assume we are given natural numbers

$$1, 2, \ldots, S \geqslant |a_0| = |r_1||r_2|\ldots|r_m| \geqslant |r_m|, \tag{14.1.2}$$

where $|a_0|$ is the absolute value of the constant and $|r_m|$ is the largest absolute value of the roots. Suppose that $N = 2S$. Here the problem is of searching the roots of the function by a query. One possible solution can be of determining the following values

$$f(-S), \ldots, f(-2), f(-1), f(1), f(2), \ldots, f(S). \tag{14.1.3}$$

By considering the classical case, 2^N queries are needed, which are 2^N separate evaluations of the values of the function (14.1.1). In the suggested quantum algorithm, the main aim is using a single query. Suppose g as another function defined as

$$g: \mathbf{R} \to \{0, 1\}. \tag{14.1.4}$$

doi:10.1088/978-0-7503-4700-6ch14

The relation between the functions $f(x)$ and $g(x)$ is defined as follows:

$$g(x) = \begin{cases} 1 & \text{if } f(x) \neq 0, \\ 0 & \text{if } f(x) = 0. \end{cases} \tag{14.1.5}$$

Our aim is of finding the values x by which $f(x) = 0$ since they would be the roots of $f(x)$. According to the relation, by finding the values x such that $g(x) = 0$, the roots of $f(x)$ can be determined. One step is, therefore, of determining the following N values (a bit string)

$$g(-S), \dots, g(-2), g(-1), g(1), g(2), \dots, g(S). \tag{14.1.6}$$

The Bernstein–Vazirani algorithm generalized to the modulo 2 is utilized for determining a bit string. Suppose that

$$\overbrace{g(-S), \dots, g(S)}^{N} = 0, 1. \tag{14.1.7}$$

The following quantum state

$$|\vec{2} - (g(-S), \dots, g(S))\rangle \tag{14.1.8}$$

can be obtained. That is to say, if we measure $|\vec{2} - (g(-S), \dots, g(S))\rangle$ then we can retrieve the following bit string

$$g(-S), \dots, g(S) \tag{14.1.9}$$

using a single query. If $g(-3) = 0, g(5) = 0, \dots, g(S-2) = 0$, the function $f(x)$ can be factorized as follows:

$$f(x) = (x+3)(x-5)\dots(x-(S-2)). \tag{14.1.10}$$

Therefore, we can find the m roots of the function. All we have to do is perform one quantum measurement. The step to determine the m roots improves by a factor of 2^N as compared to the classical counterpart.

14.2 Finding the roots of a polynomial function by using a natural number string

Now in this section, a more effective method for determining the roots of a polynomial function is investigated. The Bernstein–Vazirani algorithm generalized is used, in the modulo d, in order to figure out a natural number string.

First, consider $f(x)$ as a polynomial function defined as follows:

$$f(x) = x^m + a_{m-1}x^{m-1} + \dots + a_1 x + a_0. \tag{14.2.1}$$

Here all the roots are in the integers $\mathbf{Z}$.

Assume we are given natural numbers

$$1, 2, \dots, S \geqslant |a_0| = |r_1||r_2| \dots |r_m| \geqslant |r_m|, \tag{14.2.2}$$

where $|a_0|$ is the absolute value of the constant and $|r_m|$ is the largest absolute value of the roots. Let us assume $N = 2S$. So, the problem would be trying to find the roots of the function by one query. One step is of determining the following positive integer

$$|f(-S)|, \dots, |f(-2)|, |f(-1)|, |f(1)|, |f(2)|, \dots, |f(S)|. \tag{14.2.3}$$

We use the Bernstein–Vazirani algorithm generalized to the modulo d for determining a natural number string

$$0 \leqslant \overbrace{|f(-S)|, \dots, |f(S)|}^{N} \leqslant d - 1. \tag{14.2.4}$$

The following quantum state

$$|\vec{d} - (|f(-S)|, \dots, |f(S)|)\rangle \tag{14.2.5}$$

can be obtained. That is to say, if we measure $|\vec{d} - (|f(-S)|, \dots, |f(S)|)\rangle$ then we can retrieve the following natural numbers

$$|f(-S)|, \dots, |f(S)| \tag{14.2.6}$$

using a single query. If $|f(-3)| = 0$, $|f(5)| = 0, \dots, |f(S-2)| = 0$, the function $f(x)$ can be factorized as follows

$$f(x) = (x + 3)(x - 5) \dots (x - (S - 2)). \tag{14.2.7}$$

Therefore, we can find the m roots of the function. All we have to do is perform one quantum measurement. The speed to determine the m roots improves by a factor of d^N as compared to the classical counterpart. We find the problem becomes more easy than it using a bit string.

14.3 Finding the roots of a polynomial function by using an integer string

In this section, our third method for extracting the roots of a polynomial function is investigated. Similar to the two prior approaches, this method is based on the Bernstein–Vazirani algorithm generalized to the modulo d, by the aim of finding an integer string.

The function $f(x)$ can be shown as $f(x) = x^m + a_{m-1}x^{m-1} + \dots + a_1x + a_0$ in which all the roots are in the integers $\mathbf{Z}$.

The natural numbers $1, 2, \dots, S \geqslant |a_0| = |r_1||r_2|\dots|r_m| \geqslant |r_m|$ are taken into account. $|a_0|$ is the absolute value of the constant and $|r_m|$ is the largest absolute value of the roots.

First assume that $N = 2S$. Utilizing just one query in order to find the roots of the function is the main goal. Then, we need to determine the following sequence

$$f(-S), \dots, f(-2), f(-1), f(1), f(2), \dots, f(S). \tag{14.3.1}$$

It is worth pointing out that solving such a problem by using the classical algorithms takes $(2d - 1)^N$ queries. Suppose that

$$-(d-1) \leqslant \overbrace{f(-S), \ldots, f(S)}^{N} \leqslant d-1. \tag{14.3.2}$$

The following quantum state

$$|-(f(-S), \ldots, f(S))\rangle \tag{14.3.3}$$

can be obtained. If we measure $|-(f(-S), \ldots, f(S))\rangle$ then we can retrieve the following integers using a single query

$$f(-S), \ldots, f(S). \tag{14.3.4}$$

If $f(-3) = 0, f(5) = 0, \ldots, f(S-2) = 0$, the function $f(x)$ can be factorized as follows

$$f(x) = (x+3)(x-5)\ldots(x-(S-2)). \tag{14.3.5}$$

Therefore, we can find the m roots of the function. All we have to do is perform one quantum measurement. The speed to determine the m roots improves by a factor of $(2d-1)^N$ as compared to the classical counterpart.

By having such an approach, the problem pertained to find the roots of the function can be efficiently solved and it becomes more and more easy.

Parts of this chapter have been reprinted from Nagata K, Nakamura T, Geurdes H, Batle J, Farouk A, Diep D N and Patro S K 2018 Efficient quantum algorithms of finding the roots of a polynomial function *Int. J. Theor. Phys.* **57** 2546

IOP Publishing

Simplified Quantum Computing with Applications

Koji Nagata, Do Ngoc Diep, Ahmed Farouk and Tadao Nakamura

Chapter 15

Quantum algorithm for rapidly plotting a function

In this chapter, we discuss the Bernstein–Vazirani algorithm generalized for determining an integer string. The generalized algorithm presented here has the following structure. Given the set of real values $\{a_1, a_2, a_3, \dots, a_N\}$ and a special function g, we determine simultaneously N values of the function $g(a_1), g(a_2), g(a_3), \dots, g(a_N)$. The speed of determining the strings is shown to outperform the best classical case by a factor of $(2d - 1)^N$.

15.1 Description of the algorithm

Again, this section is devoted to our introduced method to figure out an integer string. First, consider $a_1, a_2, a_3, \dots, a_N$ as a sequence of real values. The function g is defined as follows:

$$g: \mathbf{R} \to \mathbf{Z}. \tag{15.1.1}$$

Now, $g(a_1), g(a_2), g(a_3), \dots, g(a_N)$ should be determined as an integer string. To calculate this sequence by using the classical approach, $(2d - 1)^N$ queries would be utilized. What we want to achieve in this section is to accomplish this calculation just by a single query.

Consider that d is a positive integer. The problem will be considered by modulo d. Assume that the answers are between $-(d - 1)$ and $(d - 1)$ as can be seen in the following equation

$$-(d-1) \leqslant \overbrace{g(a_1), g(a_2), g(a_3), \dots, g(a_N)}^{N} \leqslant d - 1, \tag{15.1.2}$$

where $g(a_j) \in \{-(d - 1), \dots, -1, 0, 1, \dots, d - 1\}$, So, $g(a)$ can be illustrated as follows:

doi:10.1088/978-0-7503-4700-6ch15

$$g(a) = (g(a_1), g(a_2), g(a_3), \ldots, g(a_N)), \tag{15.1.3}$$

where each entry of $g(a)$ is an integer. Here $g(a) \in \{-(d-1), \ldots, -1, 0, 1, \ldots, d-1\}^N$. In the next step, consider $f(x)$ as follows:

$$f(x) = g(a) \cdot x \bmod d = g(a_1)x_1 + g(a_2)x_2 + \cdots + g(a_N)x_N \bmod d, \tag{15.1.4}$$

where $x = (x_1, \ldots, x_N) \in \{-(d-1), \ldots, -1, 0, 1, \ldots, d-1\}^N$. The next step is preparing the required quantum states. The input state is $|\psi_0\rangle$ depictured as the following equation

$$|\psi_0\rangle = |0\rangle^{\otimes N}|d-1\rangle, \tag{15.1.5}$$

where $|0\rangle^{\otimes N}$ means $\overbrace{|0,0, \ldots, 0\rangle}^{N}$. Here $|0\rangle$ is a $(2d-1)$-dimensional state and $|d-1\rangle$ is a d-dimensional state. Our aim is for obtaining the following state:

$$|-(g(a_1), g(a_2), g(a_3), \ldots, g(a_N))\rangle. \tag{15.1.6}$$

This problem can be solved using the Bernstein–Vazirani algorithm of determining an integer string. That is to say, if we measure the first N $(2d-1)$-dimensional states of the state $|\psi_3\rangle$, that is, $|-(g(a_1), g(a_2), g(a_3), \ldots, g(a_N))\rangle$, then we can retrieve the following

$$-(g(a_1), g(a_2), g(a_3), \ldots, g(a_N)) \tag{15.1.7}$$

using a single query. Thus we obtain the following integer string

$$g(a_1), g(a_2), g(a_3), \ldots, g(a_N) \tag{15.1.8}$$

using a single query. The step of determining the string is shown to outperform the classical case by a factor of $(2d-1)^N$. Therefore, by the use of the algorithm for determining an integer string, we can determine the maximum and the minimum of the function g with the discrete domain. Roughly speaking, we can plot the function g more speedily than with a classical method when $N \gg 1$.

IOP Publishing

Simplified Quantum Computing with Applications

Koji Nagata, Do Ngoc Diep, Ahmed Farouk and Tadao Nakamura

Chapter 16

Efficient exact quantum algorithm for the parity problem of a function

In this chapter, it is our aim here to introduce a new quantum algorithm that evaluates whether or not a function $f(\cdot)$ is even, i.e., $f(x) = f(-x)$. We shall evaluate the number of steps required (how fast we can reach the answer). As we shall see, all we need is to evaluate the outcome of the measurement of the state $|\overbrace{0,0,\ \ldots,\ 0,1}^{N}\rangle$, with $N \geqslant 2$. As far as the speed-up is concerned, we shall prove that the algorithm is superior to its classical counterpart by an exponential factor $O(2^N)$. Here, in chapter 16 and chapter 17, $|-x\rangle = -|x\rangle$ holds.

16.1 Description of the algorithm

Assume $|-x\rangle = -|x\rangle$. This is realized by using the following local unitary operations:

- Prepare $|-x\rangle$.
- Introduce the flip operator $\sigma_x = |-x\rangle\langle x| + |x\rangle\langle -x|$.
- Notice $\sigma_x|-x\rangle = |x\rangle$.
- Operate $-I$ to $|x\rangle$ in giving $-|x\rangle$.

Notice

- Prepare $-|x\rangle$.
- Introduce the flip operator $\sigma_x = |-x\rangle\langle x| + |x\rangle\langle -x|$.
- Notice $\sigma_x(-|x\rangle) = -| - x\rangle$.
- Operate $-I$ to $-|-x\rangle$ in giving $|-x\rangle$.

Therefore, transformations $|-x\rangle$ to $-|x\rangle$ and $-|x\rangle$ to $|-x\rangle$ are realized by using local unitary operations. In the following, local unitary operations can be used in order to justify the assumption $|-x\rangle = -|x\rangle$. Roughly speaking, all the argumentations in this section are held under the assumption that local unitary operations can be used.

doi:10.1088/978-0-7503-4700-6ch16

Let us suppose that we are given the following function

$$f: \{-(2^N - 1), -(2^N - 2), \dots, 2^N - 2, 2^N - 1\} \\ \to \{0, 1, \dots, 2^N - 2, 2^N - 1\}. \tag{16.1.1}$$

We shall assume that $N \geqslant 2$ and $f(y) \geqslant 0$. We aim to determine whether $f(-y) = f(y)$, i.e., to check the parity of the function. Classically, we have to evaluate 2^N terms of the function to ascertain its parity, i.e., we need 2^N calls or steps. However, we shall see that we require just a single query in the quantum case presented here.

Let us introduce a function $g(x)$ that transforms a binary string into an integer. We also define $g^{-1}(f(g(x))) = F(x)$. We shall assume, for the time being, that the given function is even. Thus, we have

$$F(x) = F(-x) \in \{0, 1\}^N, \tag{16.1.2}$$

where $x \in \{0, 1\}^N$. We will see that the condition (16.1.2) holds in the quantum algorithm.

What the function $f(x)$ does in (16.1.1) is to map a set of discrete values onto another one. In (16.1.2), we assume that x is the binary representation of one element. x is given by a binary string belonging to the Cartesian product $\overbrace{\{0,1\} \times \{0,1\} \times \dots \times \{0,1\}}^{N}$, for instance, $x = (0, 1, 1, 0, 0, 1, \dots, 1)$. We then define $-x$ as $-(0, 1, 1, 0, 0, 1, \dots, 1)$. The basic ingredient that we will employ to obtain a quantum speed-up is the superposition principle. Let us follow the evolution of a quantum state through the main algorithm. Throughout the discussion, we will omit any normalization factor. Let us define $|-x\rangle = -|x\rangle$. The input state is

$$|\psi_1\rangle = |\overbrace{0,0, \dots, 0,1}^{N}\rangle|\overbrace{1,1, \dots, 1}^{N}\rangle. \tag{16.1.3}$$

The state $|\psi_1\rangle$ (16.1.3) can be decomposed as follows

$$\sum_{x=-(2^N-1)}^{-2} |x\rangle|\overbrace{1,1, \dots, 1}^{N}\rangle + \sum_{x=+1}^{2^N-1} |x\rangle|\overbrace{1,1, \dots, 1}^{N}\rangle. \tag{16.1.4}$$

Now, $y = g(x)$ is the integer representation of the binary string x. For example, if $y = 3$, then we have $x = (1, 1)$.

The function F is evaluated using the following unitary $2N$-qubit gate

$$U_F|x, z\rangle = |x, z + F(x)\rangle \tag{16.1.5}$$

with

$$\begin{aligned} U_F|x, z\rangle &= |x, z + F(x)\rangle \\ &\Leftrightarrow U_F(-|x, z\rangle) = -|x, z + F(x)\rangle \\ &\Leftrightarrow U_F|-x, z\rangle = |-x, z + F(x)\rangle \\ &\Leftrightarrow U_F|-x, z\rangle = |-x, z + F(-x)\rangle \end{aligned} \tag{16.1.6}$$

and employing the fact that $F(x) = F(-x)$. Here, $z + F(x) = (z_1 \oplus F_1(x), z_2 \oplus F_2(x), \ldots, z_N \oplus F_N(x))$ (the symbol $\oplus$ indicates addition modulo 2).

Let us start the main algorithm from the following

$$U_F|\psi_1\rangle = |\psi_2\rangle = \sum_{x=-(2^N-1)}^{-2} |x\rangle|F(\bar{x})\rangle + \sum_{x=+1}^{2^N-1} |x\rangle|\overline{F(x)}\rangle. \tag{16.1.7}$$

Here, for example, if $F(0, 0, \ldots, 0, 1) = (0, 1, 1, 0, 0, 1, \ldots, 1)$, then $\overline{F(0, 0, \ldots, 0, 1)} = (1, 0, 0, 1, 1, 0, \ldots, 0)$. Hence we have

$$\begin{aligned} |\psi_2\rangle &= \sum_{x=-(2^N-1)}^{-2} |x\rangle|\overline{F(x)}\rangle + \sum_{x=+1}^{2^N-1} |x\rangle|\overline{F(x)}\rangle \\ &= \sum_{x=+2}^{2^N-1} |-x\rangle|\overline{F(-x)}\rangle + \sum_{x=+1}^{2^N-1} |x\rangle|\overline{F(x)}\rangle \\ &= \sum_{x=+2}^{2^N-1} |-x\rangle|\overline{F(x)}\rangle + \sum_{x=+1}^{2^N-1} |x\rangle|\overline{F(x)}\rangle \\ &= |0, 0, \ldots, 0, 1\rangle|\overline{F(0, 0, \ldots, 0, 1)}\rangle, \end{aligned} \tag{16.1.8}$$

where $\overline{F(x)} = \overline{F(-x)}$.

Given the decomposition of $|\psi_1\rangle$ as (16.1.4), in (16.1.7) we apply the unitary gate U_F and obtain $|\psi_2\rangle$. What the relation (16.1.8) implies is that the unitary gate U_F does indeed work and the function is even. If the function is not even or the unitary gate U_F does not work, then we shall not get the result (16.1.8).

Finally, we shall measure the first N-qubit states of the state $|\psi_2\rangle$. If we obtain the state $|0, 0, \ldots, 0, 1\rangle$, then we can retrieve the parity of the function

$$\overline{F(x)} = \overline{F(-x)} \tag{16.1.9}$$

or, equivalently

$$F(x) = F(-x). \tag{16.1.10}$$

If this is the case, then our algorithm works, because we can check the condition (16.1.2) in a straightforward manner. All we have to do is perform a single quantum measurement.

Let us suppose that the outcome of our unique measurement is different, say $|0, 0, \ldots, 1, 0\rangle$. Then, the parity of the function is not even ($F(x) \neq F(-x)$) or the unitary gate U_F does not work correctly. For example, we cannot have

$$U_F|\overbrace{0,0, \ldots, 0,1}^{N}\rangle|\overbrace{1,1, \ldots, 1}^{N}\rangle = |\overbrace{0,0, \ldots, 0,1}^{N}\rangle|\overline{F(0, 0, \ldots, 0, 1)}\rangle \tag{16.1.11}$$

when the condition $F(x) = F(-x)$ does not hold or the unitary gate U_F does not work correctly.

Let us briefly summarize the requirements of our algorithm.

- All we need to do is of measuring $|\overbrace{0,0,\ldots,0,1}^{N}\rangle$.
- Then, we can determine the parity of the function, $f(y) = f(-y)$, for numbers $y \neq 1$.
- We have to check $f(1) = f(-1)$.

If we proceed as before, our quantum algorithm is superior to its classical counterpart by a factor $O(2^N)$.

We introduce a new algorithm that determines whether or not a given function $f(x)$ is even, and we show that it is superior to its classical counterpart by a factor $O(2^N)$. In other words, we introduce a new type of algorithm that achieves an exponential speed-up over the best available classical algorithm.

Although the preparation of the input state and the construction of the concomitant global unitary gate U_F are not considered here, we believe that our results are meaningful, since they introduce a simple, new quantum algorithm that outperforms a simple classical task.

Parts of this chapter have been reprinted from Nagata K, Nakamura T, Batle J and Farouk A 2018 Efficient quantum algorithm for the parity problem of a certain function *Int. J. Theor. Phys.* **57** 3098

Chapter 17

Necessary and sufficient condition for quantum computing

In this chapter, a necessary and sufficient condition for quantum computing is proposed. Surprisingly the relation $f(x) = f(-x)$ is the necessary and sufficient condition of holding this fundamental relation if local unitary operations can be used. A quantum algorithm (computing) experiment is successful if $f(x) = f(-x)$, otherwise the experiment is not success.

17.1 Necessary and sufficient condition for quantum computing

In this section, a necessary and sufficient condition for quantum computing is proposed. Assume $|-x\rangle = -|x\rangle$. Transformations $|-x\rangle$ to $-|x\rangle$ and $-|x\rangle$ to $|-x\rangle$ are realized by using local unitary operations. In the following, local unitary operations can be used in order to justify the assumption $|-x\rangle = -|x\rangle$. All the arguments in this section are held under the assumption that local unitary operations can be used.

Assume that the following function is given

$$f\colon \{-(2^N - 1), -(2^N - 2), \ldots, 2^N - 2, 2^N - 1\} \\ \to \{0, 1, \ldots, 2^N - 2, 2^N - 1\}. \tag{17.1.1}$$

Assume that $f(y) \geqslant 0$. Introduce a function $g(x)$ that transforms binary strings into an integer. Define $g^{-1}(f(g(x))) = F(x)$. The construction of $F(x)$ is independent of the choice of the function g. Assume such a function $F(x)(=g^{-1}(f(g(x))))$ and such a function g indeed exists, for example to be precise, choose the function g such that for any string $x = (x_N, \ldots, x_0)$, $g(x) = x_N 2^N + \cdots + x_1 2^1 + x_0$. This function g is invertible and g^{-1} converts the integer $x = x_N 2^N + \cdots + x_1 2^1 + x_0$ back to the bit-string $(x_N, \ldots, x_1, x_0)$. $y = g(x)$ is the integer representation of the binary string x. For example, $x = (1, 1)$ if $y = 3$. Assume that the given function f is even. Namely,

$$F(x) = F(-x) \in \{0, 1\}^N, \tag{17.1.2}$$

where $x \in \{0, 1\}^N$. The condition (17.1.2) is a sufficient condition for quantum computing as shown below.

What the function $f(x)$ does in (17.1.1) is of mapping a set of discrete values onto another one. In (17.1.2), we assume that x is the binary representation of an integer. x will be given by a binary string belonging to the Cartesian product $\overbrace{\{0,1\} \times \{0,1\} \times \ldots \times \{0,1\}}^{N}$, for instance, $x = (0, 1, 1, 0, 0, 1, \ldots, 1)$. Define $-x$ as $-(0, 1, 1, 0, 0, 1, \ldots, 1)$.

Throughout the discussion, any normalization factors are omitted. The input state is

$$|\psi_1\rangle = |\overbrace{0,0, \ldots, 0,1}^{N}\rangle|\overbrace{1,1, \ldots, 1}^{N}\rangle. \tag{17.1.3}$$

The function F is evaluated by using the following unitary $2N$ qubit gate

$$U_F|x, z\rangle = |x, z + F(x)\rangle \tag{17.1.4}$$

with

$$\begin{aligned} U_F|x, z\rangle &= |x, z + F(x)\rangle \\ &\Leftrightarrow U_F(-|x, z\rangle) = -|x, z + F(x)\rangle \\ &\Leftrightarrow U_F|-x, z\rangle = |-x, z + F(x)\rangle \\ &\Leftrightarrow U_F|-x, z\rangle = |-x, z + F(-x)\rangle \end{aligned} \tag{17.1.5}$$

employing the fact that $F(x) = F(-x)$. Here, $z + F(x) = (z_1 \oplus F_1(x), z_2 \oplus F_2(x), \ldots, z_N \oplus F_N(x))$ (the symbol $\oplus$ indicates addition modulo 2). And, $\overline{F(x)} = (1 \oplus F_1(x), 1 \oplus F_2(x), \ldots, 1 \oplus F_N(x))$. For example, $F(0, 0, \ldots, 0, 1) = (0, 1, 1, 0, 0, 1, \ldots, 1)$ if $\overline{F(0, 0, \ldots, 0, 1)} = (1, 0, 0, 1, 1, 0, \ldots, 0)$.

The state $|\psi_1\rangle$ (17.1.3) can be decomposed as follows

$$\sum_{x=-(2^N-1)}^{-2} |x\rangle|\overbrace{1,1, \ldots, 1}^{N}\rangle + \sum_{x=+1}^{2^N-1} |x\rangle|\overbrace{1,1, \ldots, 1}^{N}\rangle. \tag{17.1.6}$$

Start the discussion from the following:

$$U_F|\psi_1\rangle = |\psi_2\rangle = \sum_{x=-(2^N-1)}^{-2} |x\rangle|\overline{F(x)}\rangle + \sum_{x=+1}^{2^N-1} |x\rangle|\overline{F(x)}\rangle \tag{17.1.7}$$

and

$$\begin{aligned} |\psi_2\rangle &= \sum_{x=-(2^N-1)}^{-2} |x\rangle|\overline{F(x)}\rangle + \sum_{x=+1}^{2^N-1} |x\rangle|\overline{F(x)}\rangle \\ &= \sum_{x=+2}^{2^N-1} |-x\rangle|\overline{F(-x)}\rangle + \sum_{x=+1}^{2^N-1} |x\rangle|\overline{F(x)}\rangle \\ &= \sum_{x=+2}^{2^N-1} |-x\rangle|\overline{F(x)}\rangle + \sum_{x=+1}^{2^N-1} |x\rangle|\overline{F(x)}\rangle \\ &= |0, 0, \ldots, 0, 1\rangle|\overline{F(0, 0, \ldots, 0, 1)}\rangle, \end{aligned} \tag{17.1.8}$$

where $\overline{F(x)} = \overline{F(-x)}$.

The following fundamental relation in quantum computing is derived:

$$U_F|\overbrace{0,0,\ \ldots,\ 0,1}^{N}\rangle|\overbrace{1,1,\ \ldots,\ 1}^{N}\rangle = |\overbrace{0,0,\ \ldots,\ 0,1}^{N}\rangle|\overline{F(0,\ 0,\ \ldots,\ 0,\ 1)}\rangle. \tag{17.1.9}$$

Therefore, the relation $F(x) = F(-x)$ is a sufficient condition for the fundamental relation (17.1.9). The relation $F(x) = F(-x)$ is also a necessary condition for the fundamental relation (17.1.9) as shown below.

From the definition in (17.1.5), notice

$$U_F|x\rangle|\overbrace{1,1,\ \ldots,\ 1}^{N}\rangle = |x\rangle|\overline{F(x)}\rangle. \tag{17.1.10}$$

This implies for $x \to -x$, with $x \neq 0$

$$U_F|-x\rangle|\overbrace{1,1,\ \ldots,\ 1}^{N}\rangle = |-x\rangle|\overline{F(-x)}\rangle. \tag{17.1.11}$$

Assume $|-x\rangle = -|x\rangle$. It follows that the minus sign on the left and right hand sides of (17.1.11) drops off. This implies

$$U_F|x\rangle|\overbrace{1,1,\ \ldots,\ 1}^{N}\rangle = |x\rangle|\overline{F(-x)}\rangle. \tag{17.1.12}$$

Assume such that

$$|P\rangle = |Q\rangle \Leftrightarrow P = Q. \tag{17.1.13}$$

Comparing (17.1.10) with (17.1.12), notice $|\overline{F(x)}\rangle = |\overline{F(-x)}\rangle$. Hence, the following property of the function in order to maintain a consistency for the fundamental relation (17.1.9) cannot be avoided:

$$\overline{F(x)} = \overline{F(-x)}. \tag{17.1.14}$$

The fact that the function under study is even is derived:

$$F(x) = F(-x). \tag{17.1.15}$$

Thus the relation $F(x) = F(-x)$ is a necessary condition for the fundamental relation (17.1.9).

It is indeed a surprise to show the relation $f(x) = f(-x)$ is the necessary and sufficient condition of holding this fundamental relation if local unitary operations can be used. Again, a quantum algorithm (computing) experiment is successful if $f(x) = f(-x)$, otherwise the experiment is not a success. It is the all-versus-nothing theorem.

A quantum algorithm does not distinguish $f(-x)$ from $f(x)$. In other words, the minus sign does not change anything in the outcome of the quantum algorithm. This fact is due to the Galilean transformation that changes the Cartesian coordinate from x to $-x$. The non-relativistic quantum theory is invariant under the Galilean transformation.

The all-versus-nothing theorem is explained as the quantum algorithms are invariant under the Galilean transformation.

There are many researches concerning quantum computing, quantum algorithm, and their experiments. However, a complete understanding of a fundamental structure of quantum computing is not given. There is a meaningful motivation of looking for a condition of obtaining the success of a quantum computing experiment. Namely, it is useful for experimental investigations for judging if the experiments are successful. For example, the Bernstein–Vazirani algorithm in a noisy environment is studied and a success probability is given. Here an all-versus-nothing theorem is investigated.

Assume a $2N$ qubit-quantum computing which starts with the state $|\overbrace{0,0,\ldots,0,1}^{N}\rangle|\overbrace{1,1,\ldots,1}^{N}\rangle$ as follows:

$$U_f|0, 0, \ldots, 0, 1\rangle|1, 1, \ldots, 1\rangle = |0, 0, \ldots, 0, 1\rangle|\overline{f(0, 0, \ldots, 0, 1)}\rangle. \quad (17.1.16)$$

Recently, it was shown that the relation $f(x) = f(-x)$ is a necessary condition of holding this fundamental relation of quantum computing performed with, for example, the Deutsch–Jozsa algorithm or the Bernstein–Vazirani algorithm if we assume $|-x\rangle = -|x\rangle$. It is interesting to study whether the relation $f(x) = f(-x)$ is also a sufficient condition.

The argumentations for finding this evenness: $f(x) = f(-x)$ of the function $f(x)$ is assumed to a practical use in combination together with quantum error correction algorithms for qubits/qudits (or even topological anyons in topological quantum error correction). After an error correction the problem is protected from any kinds of bit flip or spin flip actions and the argumentations work well.

Parts of this chapter have been reprinted from Nagata K, Nakamura T, Farouk A and Diep D N 2019 Necessary and sufficient condition for quantum computing *Int. J. Theor. Phys.* **58** 136

IOP Publishing

Simplified Quantum Computing with Applications

Koji Nagata, Do Ngoc Diep, Ahmed Farouk and Tadao Nakamura

Chapter 18

Toward practical quantum-gated computers

In this chapter, we discuss the concept of a very true quantum computing that evolves from physics evolution based upon quantum mechanics. We show various multiple operations through both the superposition and the phase factor connected by the phase kickback formation, in order to store the routine of a Boolean function used to compute mathematical computations in the binary system. We review the way of storing, in the quantum-gated computer, logical functions in a Boolean algebra, that was developed internationally first by Nagata and Nakamura. And then, we demonstrate typical arithmetic calculations in the binary system. Thus we may lead the way to implementing commercial quantum-gated computers.

18.1 Quantum algorithm for storing all the mappings of a logical function

We show various multiple operations through both the superposition and the phase factor connected by the phase kickback formation, in order to store the routine of a Boolean function used to compute mathematical computations in the binary system.

Let us consider a logical function which is constructed based on a Boolean algebra stemming from atoms in it: there are the 2^{2^2} functions $(f_0, f_1, \dots, f_{15})$ in a Boolean algebra F_2. Thus, there are the 2^{2^2} mappings U_{f_i}. Here, we consider f_1:

$$f_1(x, y) = A \wedge B, \tag{18.1.1}$$

where x and y are variables used in mapping f. $x(=0, 1)$ is the variable for A. $y(=0, 1)$ is the variable for B. We want to store simultaneously all the 2^2 mappings of the Boolean function:

$$f_1(00) = 0, f_1(01) = 0, f_1(10) = 0, f_1(11) = 1, \tag{18.1.2}$$

where we make a combination between a unitary transformation theory and logic theory. That is, the abbreviations $f_1(00), f_1(01), f_1(10), f_1(11)$ stand for $f_1(0, 0)$,

doi:10.1088/978-0-7503-4700-6ch18

$f_1(0, 1)$, $f_1(1, 0)$, $f_1(1, 1)$, respectively. In the classical case we require 2^{2^2} steps for storing all the 2^2 mappings. In the quantum case we require just one step.

We define the following notations:

$$|-\rangle_y = \frac{|0\rangle - i|1\rangle}{\sqrt{2}}, \ |+\rangle_y = \frac{|0\rangle + i|1\rangle}{\sqrt{2}}, \ |-\rangle_x = \frac{|0\rangle - |1\rangle}{\sqrt{2}}. \tag{18.1.3}$$

We define the input state as follows:

$$\begin{gathered} |\psi_0\rangle = a_1|00\rangle|-\rangle_y + a_2|01\rangle|-\rangle_y + a_3|10\rangle|-\rangle_x + a_4|11\rangle|-\rangle_x, \\ \langle\psi_0|\psi_0\rangle = 1 \Leftarrow |a_1|^2 + |a_2|^2 + |a_3|^2 + |a_4|^2 = 1, \\ a_1 \neq 0, a_2 \neq 0, a_3 \neq 0, a_4 \neq 0. \end{gathered} \tag{18.1.4}$$

Before the phase kickback formation, we have

$$\begin{aligned} U_{f_1}|\psi_0\rangle = a_1|00\rangle&\frac{|0 \oplus f_1(00)\rangle - i|1 \oplus f_1(00)\rangle}{\sqrt{2}} \\ + a_2|01\rangle&\frac{|0 \oplus f_1(01)\rangle - i|1 \oplus f_1(01)\rangle}{\sqrt{2}} \\ + a_3|10\rangle&\frac{|0 \oplus f_1(10)\rangle - |1 \oplus f_1(10)\rangle}{\sqrt{2}} \\ + a_4|11\rangle&\frac{|0 \oplus f_1(11)\rangle - |1 \oplus f_1(11)\rangle}{\sqrt{2}}, \end{aligned} \tag{18.1.5}$$

where $\oplus$ indicates addition modulo 2. Hence, there are $f_1(00)$, $f_1(01)$, $f_1(10)$, $f_1(11)$ in the superposition state $U_{f_1}|\psi_0\rangle$.

We store simultaneously $f_1(00)$, $f_1(01)$, $f_1(10)$, $f_1(11)$ as the phase factor after the phase kickback formation. From the map U_{f_1}, we can define the following formulas (the phase kickback formation):

$$U_{f_1}|00\rangle|-\rangle_y = \begin{cases} (-i)^{f_1(00)}|00\rangle|-\rangle_y & \text{if } f_1(00) = 0, \\ (-i)^{f_1(00)}|00\rangle|+\rangle_y & \text{if } f_1(00) = 1. \end{cases} \tag{18.1.6}$$

$$U_{f_1}|01\rangle|-\rangle_y = \begin{cases} (-i)^{f_1(01)}|01\rangle|-\rangle_y & \text{if } f_1(01) = 0, \\ (-i)^{f_1(01)}|01\rangle|+\rangle_y & \text{if } f_1(01) = 1. \end{cases} \tag{18.1.7}$$

$$U_{f_1}|10\rangle|-\rangle_x = \begin{cases} (-1)^{f_1(10)}|10\rangle|-\rangle_x & \text{if } f_1(10) = 0, \\ (-1)^{f_1(10)}|10\rangle|-\rangle_x & \text{if } f_1(10) = 1. \end{cases} \tag{18.1.8}$$

$$U_{f_1}|11\rangle|-\rangle_x = \begin{cases} (-1)^{f_1(11)}|11\rangle|-\rangle_x & \text{if } f_1(11) = 0, \\ (-1)^{f_1(11)}|11\rangle|-\rangle_x & \text{if } f_1(11) = 1. \end{cases} \tag{18.1.9}$$

There are $f_1(00)$, $f_1(01)$, $f_1(10)$, $f_1(11)$ in the phase factor. They are the values of the mappings we want to store simultaneously. After the phase kickback formation, the output state is defined as follows:

$$\begin{aligned} U_{f_1}|\psi_0\rangle = & \, a_1(-i)^{f_1(00)}|00\rangle|-\rangle_y \\ & + a_2(-i)^{f_1(01)}|01\rangle|-\rangle_y \\ & + a_3(-1)^{f_1(10)}|10\rangle|-\rangle_x \\ & + a_4(-1)^{f_1(11)}|11\rangle|-\rangle_x. \end{aligned} \tag{18.1.10}$$

There are $f_1(00), f_1(01), f_1(10), f_1(11)$ in the phase factor after the phase kickback formation. Our aim is of storing simultaneously the values of the mappings. It is possible for our well-structured algorithm.

By recalling the phase factor of $U_{f_1}|\psi_0\rangle$, we store simultaneously all the 2^2 mappings of a logical function $f_1(x, y) = A \wedge B$ for all x and y:

$$f_1(00) = 0, f_1(01) = 0, f_1(10) = 0, f_1(11) = 1. \tag{18.1.11}$$

This is very interesting indeed: the quantum algorithm gives us the ability to store a perfect property of the logical function $f_1(x, y)$, namely, $f_1(x, y)$ itself. This is faster than a classical apparatus which would require at least 2^{2^2} steps.

Applying U_{f_i}, $(i = 0, 1, 2, \ldots, 2^{2^2} - 1)$ to $|\psi_0\rangle$, $U_{f_i}|\psi_0\rangle = |\psi_1\rangle_i$, therefore leaves us with one of 2^{2^2} cases.

$$\begin{aligned} &|\psi_1\rangle_0 = a_1|00\rangle|-\rangle_y + a_2|01\rangle|-\rangle_y + a_3|10\rangle|-\rangle_x + a_4|11\rangle|-\rangle_x \\ &\text{iff } f_0(00) = 0, f_0(01) = 0, f_0(10) = 0, f_0(11) = 0, \end{aligned} \tag{18.1.12}$$

$$\begin{aligned} &|\psi_1\rangle_1 = a_1|00\rangle|-\rangle_y + a_2|01\rangle|-\rangle_y + a_3|10\rangle|-\rangle_x - a_4|11\rangle|-\rangle_x \\ &\text{iff } f_1(00) = 0, f_1(01) = 0, f_1(10) = 0, f_1(11) = 1, \end{aligned} \tag{18.1.13}$$

$$\begin{aligned} &|\psi_1\rangle_2 = a_1|00\rangle|-\rangle_y + a_2|01\rangle|-\rangle_y - a_3|10\rangle|-\rangle_x + a_4|11\rangle|-\rangle_x \\ &\text{iff } f_2(00) = 0, f_2(01) = 0, f_2(10) = 1, f_2(11) = 0, \end{aligned} \tag{18.1.14}$$

$$\begin{aligned} &|\psi_1\rangle_3 = a_1|00\rangle|-\rangle_y + a_2|01\rangle|-\rangle_y - a_3|10\rangle|-\rangle_x - a_4|11\rangle|-\rangle_x \\ &\text{iff } f_3(00) = 0, f_3(01) = 0, f_3(10) = 1, f_3(11) = 1, \end{aligned} \tag{18.1.15}$$

$$\begin{aligned} &|\psi_1\rangle_4 = a_1|00\rangle|-\rangle_y - ia_2|01\rangle|+\rangle_y + a_3|10\rangle|-\rangle_x + a_4|11\rangle|-\rangle_x \\ &\text{iff } f_4(00) = 0, f_4(01) = 1, f_4(10) = 0, f_4(11) = 0, \end{aligned} \tag{18.1.16}$$

$$\begin{aligned} &|\psi_1\rangle_5 = a_1|00\rangle|-\rangle_y - ia_2|01\rangle|+\rangle_y + a_3|10\rangle|-\rangle_x - a_4|11\rangle|-\rangle_x \\ &\text{iff } f_5(00) = 0, f_5(01) = 1, f_5(10) = 0, f_5(11) = 1, \end{aligned} \tag{18.1.17}$$

$$\begin{aligned} &|\psi_1\rangle_6 = a_1|00\rangle|-\rangle_y - ia_2|01\rangle|+\rangle_y - a_3|10\rangle|-\rangle_x + a_4|11\rangle|-\rangle_x \\ &\text{iff } f_6(00) = 0, f_6(01) = 1, f_6(10) = 1, f_6(11) = 0, \end{aligned} \tag{18.1.18}$$

$$\begin{aligned} &|\psi_1\rangle_7 = a_1|00\rangle|-\rangle_y - ia_2|01\rangle|+\rangle_y - a_3|10\rangle|-\rangle_x - a_4|11\rangle|-\rangle_x \\ &\text{iff } f_7(00) = 0, f_7(01) = 1, f_7(10) = 1, f_7(11) = 1, \end{aligned} \tag{18.1.19}$$

and so on. Finally, we have

$$\begin{aligned} |\psi_1\rangle_{15} &= -ia_1|00\rangle|+\rangle_y - ia_2|01\rangle|+\rangle_y - a_3|10\rangle|-\rangle_x - a_4|11\rangle|-\rangle_x \\ \text{iff } & f_{15}(00) = 1, f_{15}(01) = 1, f_{15}(10) = 1, f_{15}(11) = 1, \end{aligned} \tag{18.1.20}$$

From (18.1.13), we have

$$\begin{aligned} |\psi_1\rangle_1 &= a_1|00\rangle|-\rangle_y + a_2|01\rangle|-\rangle_y + a_3|10\rangle|-\rangle_x - a_4|11\rangle|-\rangle_x \\ \text{iff } & f_1(00) = 0, f_1(01) = 0, f_1(10) = 0, f_1(11) = 1. \end{aligned} \tag{18.1.21}$$

Hence, we store the mappings of the logical function:

$$f_1(x, y) = A \wedge B. \tag{18.1.22}$$

From (18.1.18), we have

$$\begin{aligned} |\psi_1\rangle_6 &= a_1|00\rangle|-\rangle_y - ia_2|01\rangle|+\rangle_y - a_3|10\rangle|-\rangle_x + a_4|11\rangle|-\rangle_x \\ \text{iff } & f_6(00) = 0, f_6(01) = 1, f_6(10) = 1, f_6(11) = 0. \end{aligned} \tag{18.1.23}$$

Hence, we store the mappings of the logical function:

$$f_6(x, y) = \text{Exclusive OR}(A, B). \tag{18.1.24}$$

From (18.1.19), we have

$$\begin{aligned} |\psi_1\rangle_7 &= a_1|00\rangle|-\rangle_y - ia_2|01\rangle|+\rangle_y - a_3|10\rangle|-\rangle_x - a_4|11\rangle|-\rangle_x \\ \text{iff } & f_7(00) = 0, f_7(01) = 1, f_7(10) = 1, f_7(11) = 1. \end{aligned} \tag{18.1.25}$$

Hence, we store the mappings of the logical function:

$$f_7(x, y) = A \vee B. \tag{18.1.26}$$

We have tried to demonstrate some parallel evaluations of a two-variable function which is an element of a Boolean algebra composed of the four-atom set. The reason why we have introduced a Boolean algebra has been because quantum gating computing with qubit systems is fundamentally using only both 0 and 1. Then quantum computing has had an ability to show some mathematically parallel computing in a Boolean algebra. We have confirmed that in quantum gating computers some algorithms can have an ability that well-structured algorithms based on Boolean algebras can have mathematically parallel evaluations beyond von Neumann architecture in terms of processing time.

In 2020, a parallel computation for all of the combinations of values in variables of a logical function was proposed by Nagata and Nakamura. The concrete quantum circuits for addition of any two numbers are proposed. In preparing all the augmentations for addition, we can expand any operation in mathematics. We expect its application for the numerical computation. We lead the way to implementing commercial quantum-gated computers.

In order to demonstrate the fact that a quantum gate method can reveal a very true parallelism used so far by von Neumann style computers, we challenge a new algorithm for evaluating a logical function which is constructed based on a Boolean algebra stemming from atoms in it. As is understandable with mathematics, quantum computing with qubit systems is included and exemplified by a Boolean algebra, which treats only both 0 and 1. Namely, the theory in this work is natural in a logical sense even in a physics domain. Therefore, quantum computing has an ability to solve some mathematical problems described in a Boolean algebra. The reason why we introduce a Boolean algebra is because we have easy and simple multiple evaluations of a function in quantum computing general. We hope the challenge here will become the initiative for the ultimate way to realize a very true parallelism on quantum computers beyond von Neumann architecture.

18.2 Toward practically mathematical evaluations

We have studied quantum operations based upon quantum mechanics. Firstly, we have used Deutsch's algorithm with the usual phase kickback formation to develop the very true overbridging between usual quantum mechanics (and then quantum computing) and a Boolean algebra. In this, we have confirmed that usual quantum operations are useful, beyond the quantum computing only for quantum mechanics operations, for very true mathematical evaluations just like an arithmetic operation. We demonstrate two typical arithmetic calculations in the binary system.

As an example of a simple addition 1 + 1 in the binary system, we are going to develop the process of how to calculate this:

To solve it, fortunately we have a formula here

$$f_6(x, y) = \text{Exclusive OR}(A, B). \tag{18.2.1}$$

$$f_1(x, y) = A \wedge B. \tag{18.2.2}$$

$$1 + 1 = ?? \tag{18.2.3}$$

$$\text{Sum} = \text{Exclusive OR}(1, 1) = 0. \tag{18.2.4}$$

$$\text{Carry} = 1 \wedge 1 = 1. \tag{18.2.5}$$

Hence we have very clearly

$$1 + 1 = 10 \tag{18.2.6}$$

according to the algorithm for addition in the binary system.

In more detail, we must use the rule of a half adder (see the figures of half adder) that is composed of by using some formulae in the Boolean algebra. In the half adder, the function of it is the SUM and a Carry to the next digit position. The circuit consists of two Boolean functions (18.2.4) and (18.2.5).

Further, we could mention a somewhat complicated example. It is 2 + 3 in the decimal system.

In addition of the half adder operation, we need one more operation the full adder (see the figures of full adder). As for the full adder, it is by the two half adders and the 'OR' operation $A \vee B$ $(A, B \in \{0, 1\})$ in the Boolean algebra to take out the result from the previous digit. The operation is left here because it is obvious mathematically. Anyhow we can obtain the result 5 in the decimal system.

To solve it, fortunately we have a formula here

$$f_6(x, y) = \text{Exclusive OR}(A, B). \tag{18.2.7}$$

$$f_1(x, y) = A \wedge B. \tag{18.2.8}$$

$$f_7(x, y) = A \vee B. \tag{18.2.9}$$

$$10 + 11 = ??? \tag{18.2.10}$$

$$\text{Sum} = \text{Exclusive OR}(0, 1) = 1. \tag{18.2.11}$$

$$\text{Carry} = 0 \wedge 1 = 0. \tag{18.2.12}$$

Thus we have

$$10 + 11 = ??1. \tag{18.2.13}$$

Also we see

$$\text{Carry } C_i = 0. \tag{18.2.14}$$

Our second aim is of calculating 1 + 1 considering Carry $C_i = 0$ by using a full adder. The first half adder says

$$\text{Exclusive OR}(1, 1) = 0. \tag{18.2.15}$$

$$\text{Carry} = 1 \wedge 1 = 1. \tag{18.2.16}$$

The second half adder says

$$\text{Sum} = \text{Exclusive OR}(\text{Carry } C_i, \text{Exclusive OR}(1, 1)) = 0. \tag{18.2.17}$$

$$\text{Carry} = \text{Carry } C_i \wedge \text{Exclusive OR}(1, 1) = 0. \tag{18.2.18}$$

Thus we see 10 + 11 =?01. We have finally the carry Carry C_0 as follows: (This is (18.2.16) ∨ (18.2.18)).

$$\text{Carry } C_0 = 0 \vee 1 = 1. \tag{18.2.19}$$

Hence we have very clearly

$$10 + 11 = 101 \tag{18.2.20}$$

according to the algorithm for addition in the binary system.

18.3 Concrete quantum circuits for addition of any two numbers

In order to practically compute the addition of any numbers, we here mention the principle, based on quantum operation and Boolean logic, of the half adder and full adder related to the addition of two numbers.

Firstly, we show the half adder for the sum of A and B at the most right digit (the last digit) in the addition of any two numbers in figure 18.1, where we have the sum

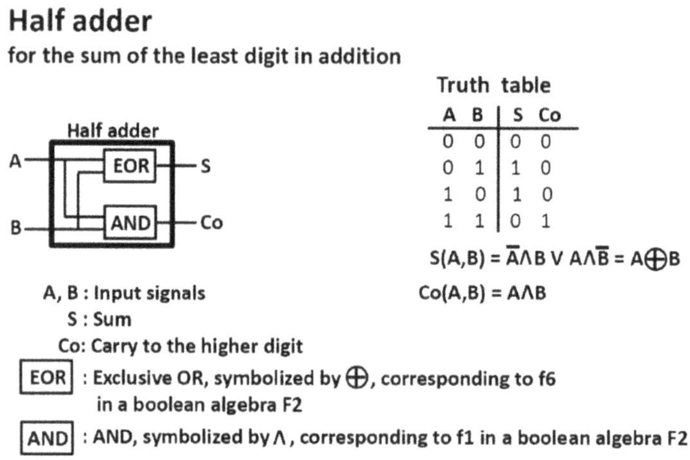

Figure 18.1. Half adder for the most right digit (the last digit) in the addition of any two numbers. Reproduced from [1], copyright 2021 with permission from Springer.

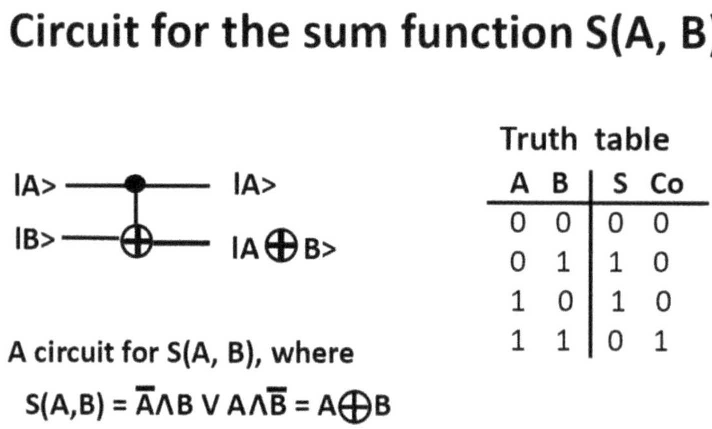

Figure 18.2. Circuit for the sum function of the two numbers at the most right digit (the last digit) in the addition. Reproduced from [1], copyright 2021 with permission from Springer.

S(A, B) of A and B and the carry $C_0(A, B)$ to the next digit. The sum S is made in figure 18.2 and the carry C_0 in figure 18.3.

Second, we also show the full adder for the sum at any digit, except at the most right digit (the last digit), in the same addition of any two numbers in figure 18.4. In this case, we have the sum $S(C_i, A, B)$ of A, B at the digit and the carry C_i from the right next, and then the carry $C_0(C_i, A, B)$ for the left next digit. The sum S is made in figure 18.5 and the carry C_0 in figure 18.6.

In preparing these elements for addition, we can expand any operation in mathematics. Therefore, the quantum computer can solve all the four basic operations of arithmetic, addition, subtraction, multiplication, and division.

Parts of this chapter have been reprinted from Nakamura T and Nagata K 2021 Physics' evolution toward computing *Int. J. Theor. Phys.* **60** 70

Circuit for the carry function Co(A, B)

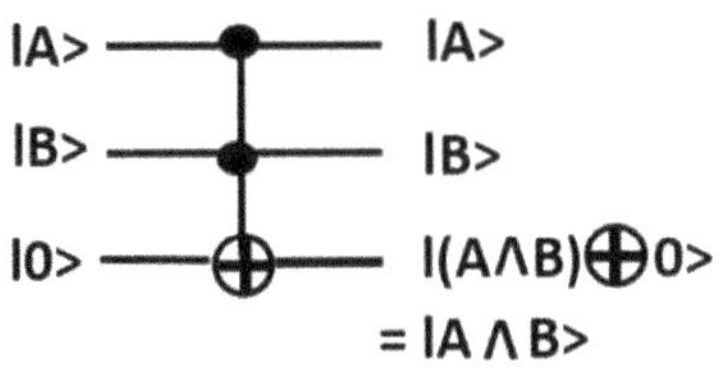

Truth table

A	B	S	Co
0	0	0	0
0	1	1	0
1	0	1	0
1	1	0	1

A circuit for Co(A, B), where

Co(A,B) = A ∧ B

Figure 18.3. Circuit for the carry function of the carry to the left next digit in the addition. Reproduced from [1], copyright 2021 with permission from Springer.

Full adder

for the sum of each digit except the least in addition

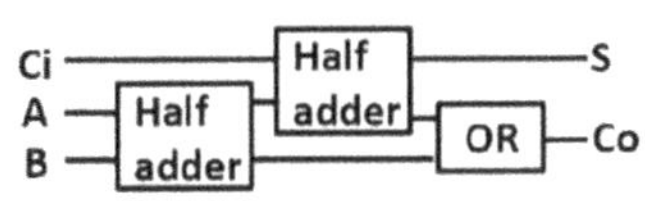

Truth table

Ci	A	B	S	Co
0	0	0	0	0
0	0	1	1	0
0	1	0	1	0
0	1	1	0	1
1	0	0	1	0
1	0	1	0	1
1	1	0	0	1
1	1	1	1	1

Ci, A, B : Input signals each of which is either "0" or "1"
(Ci is the carry signal from the lower digit)
S : Sum
Co: Carry to the higher digit

OR : OR, symbolized by V, corresponding to f7 in a boolean algebra F2

$$S(Ci, A, B) = \overline{Ci} \wedge \overline{A} \wedge B \vee \overline{Ci} \wedge A \wedge \overline{B} \vee Ci \wedge \overline{A} \wedge \overline{B} \vee Ci \wedge A \wedge B$$
$$= Ci \oplus A \oplus B$$
$$Co(Ci, A, B) = \overline{Ci} \wedge A \wedge B \vee Ci \wedge \overline{A} \wedge B \vee Ci \wedge A \wedge \overline{B} \vee Ci \wedge A \wedge B$$
$$= Ci \wedge A \vee A \wedge B \vee B \wedge Ci$$

Figure 18.4. Full adder for at any digit except the most right digit (the last digit) in the addition of any two numbers. Reproduced from [1], copyright 2021 with permission from Springer.

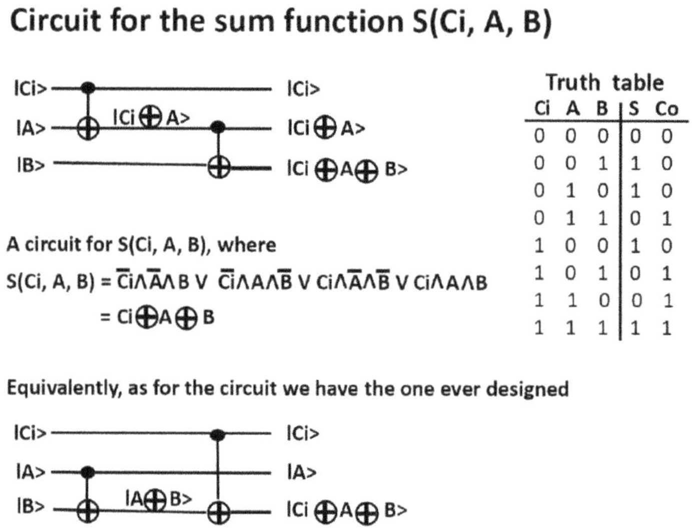

Figure 18.5. Circuits for the sum function of the two numbers at the specified digit in the addition. Reproduced from [1], copyright 2021 with permission from Springer.

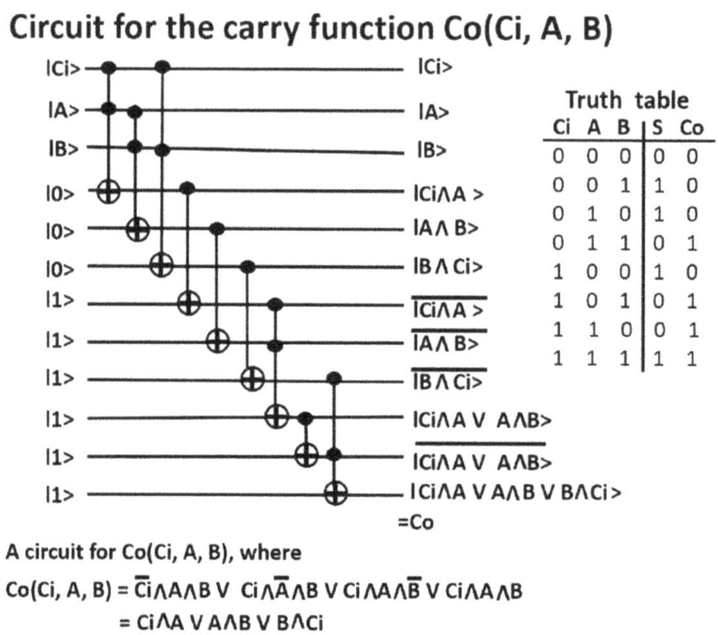

Figure 18.6. Circuit for the carry function of the carry to the left next digit in the addition. Reproduced from [1], copyright 2021 with permission from Springer.

Reference

[1] Nakamura T and Nagata K 2021 Physics' evolution toward computing *Int. J. Theor. Phys.* **60** 70

IOP Publishing

Simplified Quantum Computing with Applications

Koji Nagata, Do Ngoc Diep, Ahmed Farouk and Tadao Nakamura

Chapter 19

Computational complexity in quantum computing

In this chapter, we study an efficiency for operating a full adder/half adder by quantum-gated computing. Fortunately, we have two typical arithmetic calculations discussed by Nakamura and Nagata. The two typical arithmetic calculations are (1 + 1) and (2 + 3). We demonstrate some evaluations of three two-variable functions which are elements of a Boolean algebra composed of the four-atom set utilizing the Bernstein–Vazirani algorithm. This is faster than that of its classical apparatus which would require $2^{12} = 4096$ steps. Using the three two-variable functions evaluated here, we demonstrate a typical arithmetic calculation (2 + 3) in the binary system using the full adder operation. Surprisingly, the typical arithmetic calculation is faster than that of its classical apparatus, which would require $2^{12} = 4096$ steps when we introduce the full adder operation. Another typical arithmetic calculation (1 + 1) is faster than that of its classical apparatus which would require $2^8 = 256$ steps when we introduce only the half adder operation. The quantum advantage increases when two numbers we treat become very large.

19.1 Quantum algorithm for storing simultaneously all the mappings of three logical functions

We introduce the following three logical functions:

$$\begin{aligned} f_1(x, y) &= A \wedge B, \\ f_6(x, y) &= \text{Exclusive OR}(A, B), \\ f_7(x, y) &= A \vee B, \end{aligned} \tag{19.1.1}$$

for all x and y, that is,

$$f_1(0, 0) = 0, f_1(0, 1) = 0, f_1(1, 0) = 0, f_1(1, 1) = 1,$$
$$f_6(0, 0) = 0, f_6(0, 1) = 1, f_6(1, 0) = 1, f_6(1, 1) = 0, \tag{19.1.2}$$
$$f_7(0, 0) = 0, f_7(0, 1) = 1, f_7(1, 0) = 1, f_7(1, 1) = 1.$$

We can construct a FULL adder operation using these three logical functions.

Suppose that

$$f: \{0, 1\}^{12} \to \{0, 1\} \tag{19.1.3}$$

is a function with a 12-bit domain and a 1-bit range. We assume the following function:

$$\begin{aligned} f(x) = a \cdot x &= \sum_{i=1}^{12} a_i x_i (\text{mod } 2) \\ &= f_1(0, 0)x_1 \oplus f_1(0, 1)x_2 \oplus f_1(1, 0)x_3 \oplus f_1(1, 1)x_4 \\ &\quad \oplus f_6(0, 0)x_5 \oplus f_6(0, 1)x_6 \oplus f_6(1, 0)x_7 \oplus f_6(1, 1)x_8 \\ &\quad \oplus f_7(0, 0)x_9 \oplus f_7(0, 1)x_{10} \oplus f_7(1, 0)x_{11} \oplus f_7(1, 1)x_{12}, \end{aligned} \tag{19.1.4}$$

where $a_i \in \{0, 1\}$, $x_i \in \{0, 1\}$, $x = x_1x_2\ldots x_{12} \in \{0, 1\}^{12}$, and

$$\begin{aligned} a &= a_1a_2\ldots a_{12} \\ &= f_1(0, 0)f_1(0, 1)f_1(1, 0)f_1(1, 1)f_6(0, 0)f_6(0, 1)f_6(1, 0)f_6(1, 1) \\ &\quad \times f_7(0, 0)f_7(0, 1)f_7(1, 0)f_7(1, 1) \in \{0, 1\}^{12}. \end{aligned} \tag{19.1.5}$$

The string

$$a_1, a_2, a_3, a_4 \tag{19.1.6}$$

corresponds to

$$f_1(0, 0), f_1(0, 1), f_1(1, 0), f_1(1, 1), \tag{19.1.7}$$

respectively. The string

$$a_5, a_6, a_7, a_8 \tag{19.1.8}$$

corresponds to

$$f_6(0, 0), f_6(0, 1), f_6(1, 0), f_6(1, 1), \tag{19.1.9}$$

respectively. The string

$$a_9, a_{10}, a_{11}, a_{12} \tag{19.1.10}$$

corresponds to

$$f_7(0, 0), f_7(0, 1), f_7(1, 0), f_7(1, 1), \tag{19.1.11}$$

respectively. Here, $x \oplus y$ is the bitwise XOR (exclusive OR) of x and y. Also, $a \cdot x$ is the bitwise inner product of a and x, modulo 2. The goal is of storing the logical functions $f_1(x, y)$, $f_6(x, y)$, and $f_7(x, y)$ in a Boolean algebra for all x and y into an output quantum state as the coefficients of $f(x)$. Let us follow the quantum states through the Bernstein–Vazirani algorithm.

The input state is

$$|\psi_0\rangle = |0\rangle^{\otimes 12}|1\rangle. \tag{19.1.12}$$

After the componentwise Hadamard transformations on the state (19.1.12), we have

$$|\psi_1\rangle = \sum_{x\in\{0,1\}^{12}} \frac{|x\rangle}{\sqrt{2^{12}}}\frac{|0\rangle - |1\rangle}{\sqrt{2}}. \tag{19.1.13}$$

Next, the function f is stored into a quantum state using

$$U_f|x, y\rangle = |x, y \oplus f(x)\rangle \tag{19.1.14}$$

in giving, by the phase kickback formation,

$$|\psi_2\rangle = \pm \sum_{x\in\{0,1\}^{12}} \frac{(-1)^{f(x)}|x\rangle}{\sqrt{2^{12}}}\frac{|0\rangle - |1\rangle}{\sqrt{2}}. \tag{19.1.15}$$

After the componentwise Hadamard transformations on the first 12 qubits in the state $|\psi_2\rangle$, we can now evaluate $|\psi_3\rangle$

$$|\psi_3\rangle = \pm \sum_{z\in\{0,1\}^{12}} \sum_{x\in\{0,1\}^{12}} \frac{(-1)^{x\cdot z+f(x)}|z\rangle}{2^{12}}\frac{|0\rangle - |1\rangle}{\sqrt{2}}, \tag{19.1.16}$$

where $z_i \in \{0, 1\}$ and $z = z_1z_2\ldots z_{12} \in \{0, 1\}^{12}$. Using $f(x) = a \cdot x$, we have

$$|\psi_3\rangle = \pm \sum_{z\in\{0,1\}^{12}} \sum_{x\in\{0,1\}^{12}} \frac{(-1)^{x\cdot z+a\cdot x}|z\rangle}{2^{12}}\frac{|0\rangle - |1\rangle}{\sqrt{2}}. \tag{19.1.17}$$

Notice that

$$\sum_{x\in\{0,1\}^{12}} (-1)^{x\cdot z+a\cdot x} = 2^{12}\delta_{a,z}. \tag{19.1.18}$$

Thus, we have

$$\begin{aligned}
|\psi_3\rangle &= \pm \sum_{z\in\{0,1\}^{12}} \sum_{x\in\{0,1\}^{12}} \frac{(-1)^{x\cdot z+a\cdot x}|z\rangle}{2^{12}}\frac{|0\rangle - |1\rangle}{\sqrt{2}} \\
&= \pm \sum_{z\in\{0,1\}^{12}} \frac{2^{12}\delta_{a,z}|z\rangle}{2^{12}}\frac{|0\rangle - |1\rangle}{\sqrt{2}} \\
&= \pm\, |a_1\ldots a_{12}\rangle\frac{|0\rangle - |1\rangle}{\sqrt{2}} \\
&= \pm\, |f_1(0, 0)f_1(0, 1)f_1(1, 0)f_1(1, 1)\rangle \otimes |f_6(0, 0)f_6(0, 1)f_6(1, 0)f_6(1, 1)\rangle \\
&\quad \otimes |f_7(0, 0)f_7(0, 1)f_7(1, 0)f_7(1, 1)\rangle\frac{|0\rangle - |1\rangle}{\sqrt{2}}.
\end{aligned} \tag{19.1.19}$$

Thus, in summary, we store f_1, f_6, and f_7 into a single quantum state as

$$|f_1(0,0)f_1(0,1)f_1(1,0)f_1(1,1)\rangle \otimes |f_6(0,0)f_6(0,1)f_6(1,0)f_6(1,1)\rangle \otimes \\ |f_7(0,0)f_7(0,1)f_7(1,0)f_7(1,1)\rangle \quad (19.1.20)$$

and the results of measurements are

$$\begin{aligned} &f_1(0,0)=0, f_1(0,1)=0, f_1(1,0)=0, f_1(1,1)=1, \\ &f_6(0,0)=0, f_6(0,1)=1, f_6(1,0)=1, f_6(1,1)=0, \\ &f_7(0,0)=0, f_7(0,1)=1, f_7(1,0)=1, f_7(1,1)=1. \end{aligned} \quad (19.1.21)$$

According to the first supposition, we are succeeding in storing them correctly.

In the classical case we require $2^{12} = 4096$ steps. In the quantum case here we require just one step.

19.2 Typical arithmetic calculations

We have used the Bernstein–Vazirani algorithm with the usual phase kickback formation to develop the overbridging between usual quantum mechanics (and then quantum computing) and a Boolean algebra. In this, we have confirmed that usual quantum operations are useful, beyond the quantum computing only for quantum mechanics operations, for mathematical evaluations just like an arithmetic operation. We demonstrate two typical arithmetic calculations in the binary system.

As an example of a simple addition 1 + 1 in the binary system, we are going to develop the process of how to calculate this:

To solve it, fortunately we have a formula here

$$f_6(x, y) = \text{Exclusive OR}(A, B). \quad (19.2.1)$$

$$f_1(x, y) = A \wedge B. \quad (19.2.2)$$

$$1 + 1 = ?? \quad (19.2.3)$$

$$\text{Sum} = \text{Exclusive OR}(1, 1) = 0. \quad (19.2.4)$$

$$\text{Carry} = 1 \wedge 1 = 1. \quad (19.2.5)$$

Hence we have very clearly

$$1 + 1 = 10 \quad (19.2.6)$$

according to the algorithm for addition in the binary system. The concrete and specific calculation (1 + 1) is faster than that of a classical apparatus which would require $2^8 = 256$ steps when we introduce only the half adder operation.

In more detail, we must use the rule of a half adder that is composed by using some formulae in the Boolean algebra. In the half adder, the function of it is the SUM and a Carry to the next digit position. The circuit consists of two Boolean functions (19.2.4) and (19.2.5).

Further, we could mention a somewhat complicated example. It is 2 + 3 in the decimal system.

In addition to the half adder operation, we need one more operation, the full adder. As for the full adder, it is by the two half adders and the 'OR' operation $A \vee B$ ($A, B \in \{0, 1\}$) in the Boolean algebra to take out the result from the previous digit. The operation is left here because it is obvious mathematically. Anyhow, we can obtain the result 5 in the decimal system.

To solve it, fortunately we have a formula here

$$f_6(x, y) = \text{Exclusive OR}(A, B). \tag{19.2.7}$$

$$f_1(x, y) = A \wedge B. \tag{19.2.8}$$

$$f_7(x, y) = A \vee B. \tag{19.2.9}$$

$$10 + 11 = ??? \tag{19.2.10}$$

$$\text{Sum} = \text{Exclusive OR}(0, 1) = 1. \tag{19.2.11}$$

$$\text{Carry} = 0 \wedge 1 = 0. \tag{19.2.12}$$

Thus we have

$$10 + 11 = ??1. \tag{19.2.13}$$

Also we see

$$\text{Carry } \mathrm{C}_i = 0. \tag{19.2.14}$$

Our second aim is of calculating 1 + 1 considering Carry $\mathrm{C}_i = 0$ by using a full adder. The first half adder says

$$\text{Exclusive OR}(1, 1) = 0. \tag{19.2.15}$$

$$\text{Carry} = 1 \wedge 1 = 1. \tag{19.2.16}$$

The second half adder says

$$\text{Sum} = \text{Exclusive OR}(\text{Carry } \mathrm{C}_i, \text{Exclusive OR}(1, 1)) = 0. \tag{19.2.17}$$

$$\text{Carry} = \text{Carry } \mathrm{C}_i \wedge \text{Exclusive OR}(1, 1) = 0. \tag{19.2.18}$$

Thus we see $10 + 11 = ?01$. We have finally the carry Carry C_0 as follows: (This is (19.2.16) $\vee$ (19.2.18)).

$$\text{Carry } \mathrm{C}_0 = 0 \vee 1 = 1. \tag{19.2.19}$$

Hence we have very clearly

$$10 + 11 = 101 \tag{19.2.20}$$

according to the algorithm for addition in the binary system. The concrete and specific calculation (2 + 3) is faster than that of a classical apparatus which would require $2^{12} = 4096$ steps when we introduce the full adder operation. The quantum advantage increases when two numbers we treat become very large. Toward practical quantum-gated computers, experimental demonstrations of our argumentations are going to attract interest.

In conclusion, we have studied an efficiency for operating a full adder/half adder by quantum-gated computing. Fortunately, we have the two typical arithmetic calculations discussed by Nakamura and Nagata. We have demonstrated some evaluations of three two-variable functions which are elements of a Boolean algebra composed of the four-atom set utilizing the Bernstein–Vazirani algorithm. This has been faster than that of a classical apparatus which requires $2^{12} = 4096$ steps. Using the three two-variable functions evaluated here, we have demonstrated a typical arithmetic calculation in the binary system using the full adder operation. Surprisingly, the typical arithmetic calculation has been faster than that of a classical apparatus which requires $2^{12} = 4096$ steps when we introduce the full adder operation. Another typical arithmetic calculation has been faster than that of a classical apparatus which requires $2^8 = 256$ steps when we introduce only the half adder operation. The quantum advantage has increased when two numbers we treat become very large. Toward practical quantum-gated computers, experimental demonstrations of our argumentations are of interest.

IOP Publishing

Simplified Quantum Computing with Applications

Koji Nagata, Do Ngoc Diep, Ahmed Farouk and Tadao Nakamura

Chapter 20

Measurement theory in Deutsch's algorithm based on the truth values

In this chapter, we propose a new measurement theory, in qubits handling, based on the truth values, i.e., the truth T (1) for true and the falsity F (0) for false. The results of measurement are either 0 or 1. To implement Deutsch's algorithm, we need both observability and controllability of a quantum state. The new measurement theory can satisfy these two. Especially, we systematically describe our assertion based on more mathematical analysis using raw data in a thoughtful experiment.

20.1 The new measurement theory can satisfy observability

In this chapter, by using the double-slit experiment, we consider the relation between the new measurement theory and observability. In particular, we systematically describe our assertion based on more mathematical analysis using raw data in a thoughtful experiment.

We assume an implementation of the double-slit experiment. There is a detector just after each slit. Thus an interference image does not appear, and we do not consider such a pattern. The possible values of the result of measurements are either 1 or 0 (in $\hbar/2$ unit). If a particle passes through one side slit, then the value of the result of measurement is +1. If a particle passes through another slit, then the value of the result of measurement is 0. This is an easy detector model of a single Pauli observable.

20.2 Wave function analysis

Let σ_x be a single Pauli observable. We assume that a source of spin-carrying particles emits them in a state ρ. We consider a quantum expected value $\mathrm{Tr}[\rho\sigma_x]$. If we consider only a wave function analysis, the possible values of the square of the quantum expected value are

$$0 \leqslant (\mathrm{Tr}[\rho\sigma_x])^2 \leqslant 1. \qquad (20.2.1)$$

We define $||E_{QM}||^2$ as

$$||E_{QM}||^2 = (\mathrm{Tr}[\rho\sigma_x])^2. \tag{20.2.2}$$

Then we have

$$||E_{QM}||^2_{\min} = 0 \text{ and } ||E_{QM}||^2_{\max} = 1. \tag{20.2.3}$$

$||E_{QM}||^2_{\max}$ and $||E_{QM}||^2_{\min}$ are the maximal and minimal possible values of the product, respectively. We get $||E_{QM}||^2_{\min} = 0$ or $||E_{QM}||^2_{\max} = 1$ if the system is a pure state lying in either the z-axis or the x-axis, respectively.

20.3 New measurement theory

A mean value E satisfies the new measurement theory if it can be written as

$$E = \frac{\sum_{l=1}^{m} r_l(\sigma_x)}{m}, \tag{20.3.1}$$

where l denotes a notation and r is the result of the new measurement of the Pauli observable σ_x. We assume the values of r are either 1 or 0 (in $\hbar/2$ unit). Assume the quantum mean values with the system in a state admits the new measurement theory. One has the following proposition concerning the new measurement theory

$$\mathrm{Tr}[\rho\sigma_x](m) = \frac{\sum_{l=1}^{m} r_l(\sigma_x)}{m}. \tag{20.3.2}$$

We can assume the following by Strong Law of Large Numbers,

$$\mathrm{Tr}[\rho\sigma_x](+\infty) = \mathrm{Tr}[\rho\sigma_x]. \tag{20.3.3}$$

We define $||E_{QM}||^2(m)$ as

$$||E_{QM}||^2(m) = (\mathrm{Tr}[\rho\sigma_x](m))^2. \tag{20.3.4}$$

We can assume the following by Strong Law of Large Numbers,

$$||E_{QM}||^2(+\infty) = ||E_{QM}||^2 = (\mathrm{Tr}[\rho\sigma_x])^2. \tag{20.3.5}$$

In what follows, we show that we can accept the relation (20.3.2) concerning the new measurement theory. Assume the proposition (20.3.2) is true.

We have the following when the system is in a pure state lying in the x-axis.

$$\begin{aligned} ||E_{QM}||^2(m) &= \frac{\sum_{l=1}^{m} r_l(\sigma_x)}{m} \times \frac{\sum_{l'=1}^{m} r_{l'}(\sigma_x)}{m} \leqslant \frac{\sum_{l=1}^{m}}{m} \cdot \frac{\sum_{l'=1}^{m}}{m} |r_l(\sigma_x) r_{l'}(\sigma_x)| \\ &= \frac{\sum_{l=1}^{m}}{m} \times \frac{\sum_{l'=1}^{m}}{m} = 1. \end{aligned} \tag{20.3.6}$$

Clearly, the above inequality can have the upper limit since the following case is possible:

$$||\{l | l \in \mathbf{N} \wedge r_l(\sigma_x) = 1\}|| = ||\{l' | l' \in \mathbf{N} \wedge r_{l'}(\sigma_x) = 1\}||, \tag{20.3.7}$$

and

$$\|\{l|l \in \mathbf{N} \wedge r_l(\sigma_x) = 0\}\| = \|\{l'|l' \in \mathbf{N} \wedge r_{l'}(\sigma_x) = 0\}\|. \tag{20.3.8}$$

And we have the following when the system is in a pure state lying in the z-axis.

$$\begin{aligned} \|E_{\mathrm{QM}}\|^2(m) &= \frac{\sum_{l=1}^{m} r_l(\sigma_x)}{m} \times \frac{\sum_{l'=1}^{m} r_{l'}(\sigma_x)}{m} \geqslant \frac{\sum_{l=1}^{m}}{m} \cdot \frac{\sum_{l'=1}^{m}}{m}(0) \\ &= (0)\left(\frac{\sum_{l=1}^{m}}{m} \times \frac{\sum_{l'=1}^{m}}{m}\right) = 0. \end{aligned} \tag{20.3.9}$$

Clearly, the above inequality can have the lower limit since the following case is possible:

$$\|\{l|l \in \mathbf{N} \wedge r_l(\sigma_x) = 1\}\| = \|\{l'|l' \in \mathbf{N} \wedge r_{l'}(\sigma_x) = 0\}\|, \tag{20.3.10}$$

and

$$\|\{l|l \in \mathbf{N} \wedge r_l(\sigma_x) = 0\}\| = \|\{l'|l' \in \mathbf{N} \wedge r_{l'}(\sigma_x) = 1\}\|. \tag{20.3.11}$$

Thus we derive a proposition concerning the quantum mean value under the assumption that the new measurement theory is true (in a spin-1/2 system), that is

$$0 \leqslant \|E_{\mathrm{QM}}\|^2(m) \leqslant 1. \tag{20.3.12}$$

From Strong Law of Large Numbers, we have

$$0 \leqslant \|E_{\mathrm{QM}}\|^2 \leqslant 1. \tag{20.3.13}$$

Hence we derive the following proposition concerning the new measurement theory

$$\|E_{\mathrm{QM}}\|^2_{\min} = 0 \text{ and } \|E_{\mathrm{QM}}\|^2_{\max} = 1. \tag{20.3.14}$$

We can accept simultaneously the two relations (20.2.3) (concerning a wave function analysis) and (20.3.14) (concerning the new measurement theory). The new measurement theory meets the wave function analysis and can measure Pauli observable σ_x correctly. Similar to the argumentations, the new measurement theory can measure Pauli observable σ_z correctly. In short, the new measurement theory meets observability of σ_z and σ_x.

20.4 The new measurement theory can satisfy controllability

In this section, we investigate if the new measurement theory can satisfy controllability of a pure state lying in either the z-axis or the x-axis. These quantum states are used in Deutsch's algorithm.

We introduce the following quantum proposition concerning controllability:

$$\langle 0|0\rangle = 1,\ \langle 1|1\rangle = 1,\ \langle 0|1\rangle = 0, \text{ and } \langle 1|0\rangle = 0. \tag{20.4.1}$$

The proposition (20.4.1) implies

$$|\langle 0|0\rangle|^2 = 1,\ |\langle 1|1\rangle|^2 = 1,\ |\langle 0|1\rangle|^2 = 0,\text{ and } |\langle 1|0\rangle|^2 = 0. \tag{20.4.2}$$

Clearly, the new measurement theory based on the truth values can satisfy all the expected values in the above proposition (20.4.2). And we may assume that the new measurement theory can satisfy the proposition (20.4.1).

The proposition (20.4.1) implies that

$$\langle\sigma_z\rangle^2 + \langle\sigma_x\rangle^2 = 1 \tag{20.4.3}$$

when the system is in a pure state lying in either the z-axis or the x-axis. The reason is as follows: Assume a pure state lying in either the z-axis or the x-axis as

$$|\psi_x\rangle = e^{i\phi}\frac{|0\rangle \pm |1\rangle}{\sqrt{2}}, \tag{20.4.4}$$

and

$$|\psi_z\rangle = |0\rangle,\ |1\rangle, \tag{20.4.5}$$

where ϕ is a phase. These quantum states are used in Deutsch's algorithm. Let us write

$$\sigma_z = |0\rangle\langle 0| - |1\rangle\langle 1| \tag{20.4.6}$$

and

$$\sigma_x = |0\rangle\langle 1| + |1\rangle\langle 0| \tag{20.4.7}$$

Then we have

$$|\langle\psi_x|\sigma_z|\psi_x\rangle| = 0,\ |\langle\psi_x|\sigma_x|\psi_x\rangle| = 1. \tag{20.4.8}$$

and

$$|\langle\psi_z|\sigma_z|\psi_z\rangle| = 1,\ |\langle\psi_z|\sigma_x|\psi_z\rangle| = 0. \tag{20.4.9}$$

Therefore, we see

$$\langle\sigma_x\rangle^2 + \langle\sigma_z\rangle^2 = 1. \tag{20.4.10}$$

We thus see the proposition (20.4.1) implies the existence of either the z-axis or the x-axis in the Hilbert space formalism of the quantum theory. From the argumentations presented in section 20.3, the new measurement theory meets the relations (20.4.8) and (20.4.9). We see the new measurement theory can satisfy controllability of a pure state lying in either the z-axis or the x-axis, that is used in Deutsch's algorithm.

In conclusion, we have proposed a new measurement theory, in qubits handling, based on the truth values, i.e., the truth T (1) for true and the falsity F (0) for false. The results of measurement have been either 0 or 1. To implement Deutsch's algorithm, we need both observability and controllability of a quantum state. The new measurement theory can have satisfied these two. In particular, we systematically have described our assertion based on more mathematical analysis using raw data in a thoughtful experiment.

Our argumentations imply that we can perform the following Deutsch's algorithm:

- The control of quantum states relies on the wave functional analysis.
- The observation of quantum states relies on the new measurement theory based on the truth values.

Consistency between controllability and observability is necessary for an implementation of Deutsch's algorithm. And desired consistency is established.

Are there other measurement theories which can satisfy both controllability and observability? This is an open problem.

Parts of this chapter have been reprinted from Nagata K and Nakamura T 2016 Measurement theory in Deutsch's algorithm based on the truth values *Int. J. Theor. Phys.* **55** 3616

Chapter 21

Conclusions

In conclusion, we have presented a secure quantum key distribution based on a special Deutsch–Jozsa algorithm using Greenberger–Horne–Zeilinger states, discussed the application of the Bernstein–Vazirani algorithm to quantum communication including error correction and also studied the Bernstein–Vazirani algorithm in a noisy environment. We have proposed methods for evaluating many functions and for computing a multiplication operation, both of which use the generalized Bernstein–Vazirani algorithm together with many parallel quantum systems. We have studied quantum communication based on Simon's algorithm and generalized Deutsch's algorithm for determining all the mappings of a function. We proposed a parallel computation for all of the combinations of values in variables of a logical function, and we have demonstrated some parallel evaluations of two two-variable functions which are elements of a Boolean algebra composed of the four-atom set. Using quantum cryptography based on an algorithm for determining a function using qudit systems, we have proposed continuous-variable quantum computing and its applications to cryptography. We have presented various new forms of the Bernstein–Vazirani algorithm beyond qubit systems. We have searched an algorithm for rapidly determining a homogeneous linear function, constructed quantum algorithms for finding the roots of a polynomial function, and suggested a quantum algorithm for rapidly plotting a function. Considered an algorithm that determines whether or not a given function is even, we have proposed a necessary and sufficient condition for quantum computing. Finally, in studying our problem toward practical quantum-gated computers, we have proposed the concrete quantum circuits for addition of any two numbers and we have studied its computational complexity. Additionally, we have studied the measurement theory based on the truth values.

doi:10.1088/978-0-7503-4700-6ch21

www.ingramcontent.com/pod-product-compliance
Ingram Content Group UK Ltd.
Pitfield, Milton Keynes, MK11 3LW, UK
UKHW051850100926
8556IPUK00005B/559

9 780750 346986